DIFFUSION CLADDING OF METALS

DIFFUZIONNYE POKRYTIYA NA METALLAKH

ДИФФУЗИОННЫЕ ПОКРЫТИЯ НА МЕТАЛЛАХ

DIFFUSION CLADDING
OF METALS

Edited by
G. V. Samsonov
Director, Laboratory of Metallurgy of Rare Metals
and Refractory Compounds
Institute of Cermets and Special Alloys
Academy of Sciences of the Ukrainian SSR, Kiev

Translated from Russian

$\left(\frac{c}{b}\right)$ CONSULTANTS BUREAU · NEW YORK · 1967

Library of Congress Catalog Card Number 66-17190

ISBN-13: 978-1-4684-1565-0   e-ISBN-13: 978-1-4684-1563-6
DOI: 10.1007/978-1-4684-1563-6

The original Russian text, published for the Materials Science Institute of the Academy of Sciences of the Ukrainian SSR by Naukova Dumka in Kiev in 1965, has been corrected by the editor for the present edition.

© 1967 Consultants Bureau
A Division of Plenum Publishing Corporation
227 West 17 Street, New York, N. Y. 10011

PREFACE TO THE AMERICAN EDITION

One of the most effective methods of increasing the wear resistance, hardness, surface strength and high-temperature oxidation resistance of metals and alloys is the diffusion saturation of the surfaces by metals and nonmetals. For communicating and discussing the results of the numerous researches carried out in this field in the Department of Physicotechnical Problems of Materials Science, Academy of Sciences of the UkrSSR, a permanent Scientific Seminar was set up in 1961, which enjoys an ever-increasing popularity among specialists in this field.

The present collection contains papers read at the Third Session of this Seminar, held on September 25-28, 1963.

The compilers of the collection and the authors of the papers hope that its publication in the U. S. A. will enable American specialists to become acquainted with the main lines along which corresponding work is being conducted in the USSR. This should contribute to an exchange of scientific experience in this interesting field which is of such great practical importance.

G. V. Samsonov

PREFACE

This collection is comprised of papers relating to the diffusion saturation of metals and to coatings of refractory compounds. The papers discuss current problems in the theory and practice of the production of diffusion coatings on metallic materials. A means of classifying the methods of diffusion saturation is proposed, and a new method is described for calculating the diffusion parameters in a heterogeneous medium.

The book is intended for a wide circle of materials science specialists engaged in the production of new materials in various branches of industry.

CONTENTS

Classification of Methods of the Diffusion Saturation of the Surface of Alloys by Metals
   G. N. Dubinin.   1

Theory of Diffusion in Compounds with a Predominantly Ionic Component of the Bond
   Forces
   V. N. Konev.   7

Some Crystallographic Features of the Mechanism of Phase Transformations and
   Chemical Reactions in Solids
   V. I. Arkharov.   11

Determination of Diffusion Coefficients and Layer Thickness of the Phases in Reactive
   Diffusion
   V. T. Borisov, V. M. Golikov, and G. N. Dubinin   17

Glow-Discharge Siliconizing of Metals
   D. A. Prokoshkin, B. N. Arzamasov, and E. V. Ryabchenko   25

Vacuum Siliconizing of Refractory Metals
   V. E. Ivanov, E. P. Nechiporenko, V. I. Zmii, and V. M. Krivoruchko   29

Boroaluminizing of Iron and Steel
   G. V. Zemskov and N. G. Kaidash   35

Experience in the Application of Boriding in Tractor Construction
   N. S. Zinovich   43

Structure and Properties of Steels 20KhN3A and 17N3MA Carburized by Natural Gas
   Containing Added Ammonia
   N. G. Shul'ga, M. M. Fetisova, F. G. Krivenko, and E. M. Tyrman   49

Diffusion Saturation of Steel from a Gaseous Medium by High-Frequency
   Induction Heating
   Yu. V. Grdina and L. T. Gordeeva.   53

Phase Transformations in High-Temperature Gas Carbonitriding (Cyaniding)
   V. G. Permyakov and V. G. Tinyaev.   57

Application of Liquid Cyaniding by Potassium Ferrocyanide in Mass Production
   Ya. N. Funshtein   63

Surface Hardening of Titanium by Nitrogen and Carbon Using High-Frequency
   Induction Heating
   Yu. V. Grdina, L. T. Gordeeva, and L. T. Timonina   69

Chromizing of Steel by High-Frequency Induction Heating in a Vacuum
    G. V. Zemskov and L. K. Gushchin   73
The Use of Refractory Compounds in the Electrochemical Industry
    V. P. Basov, L. F. Kalinichenko, and A. P. Épik   77
Some Properties of Carbide and Boride Diffusion Coatings on Refractory Metals
    A. P. Épik, G. A. Bovkun, I. V. Golubchik, and L. P. Sinitsina   81

CLASSIFICATION OF METHODS OF THE DIFFUSION
SATURATION OF THE SURFACE OF ALLOYS BY METALS

G. N. Dubinin

Many different methods are currently in use for the diffusion saturation of the surfaces of parts by metals and metalloids to impart to them special physicochemical properties. The different published explanations of the saturation methods (solid, liquid, gas) do not reflect their physicochemical essentials, which impedes the study and improvement of the technology of the diffusing metal in a container, regardless of whether or not a gas phase is produced inside the container on heating. By the gas method is understood a saturation method in which the parts are situated in an environment of a gas phase only. Only one technical saturation method, saturation by metals in a vacuum, is termed the vacuum method. If, however, we consider that in the saturation of parts situated in contact with a powder mixture containing ammonium halide, the transfer of metal to the surface of the parts is brought about by the production of a gas phase in the container, it would be wrong to refer to such a method as a solid method, just as it would be wrong to refer to the method of gas carburization (carried out by packing the parts in a carburizer) as a solid method, being guided in this merely by the technical method employed in carrying out the saturation. It would also be just as incorrect to term, as is frequently done, the method of saturation in the powder of a readily volatilized metal (under vacuum conditions or without a vacuum) as a vacuum or solid method, since the transfer of metal to the treated surface occurs in this case, as investigations have shown [7], not through points of contact but through the vapor phase.

A study of the different saturation methods shows that they are related by common methodological criteria, which may also be used for their classification. Furthermore, investigations have shown that the structure and properties of the alloys depend substantially on the saturation method employed [7]. The compilation of the above-mentioned classification therefore becomes all the more necessary.

One and the same method may be carried out by different processes. For example, the essential feature of the gas method will not be provided if saturation in an atmosphere of the gas occurs under conditions such that the parts are in contact with a powder producing an atmosphere of the gas around the treated parts, or if the source of the gas is separated from the treated part. In such a case, the difference will only be in the saturation process, since in the first case we are dealing with a contact saturation process (pack process) and in the second with a noncontact saturation process. Even the saturation process employed, however, also has an influence on the results (see table).

The basis of the proposed classification is the physicochemical characteristic of the active phase (or medium) containing the diffusing element.

1

TABLE 1. Data on the Influence of the Method and Process of Saturating the Surface of Iron with Different Metals on the Structure and Properties of the Diffusion Layer, Duration 6 hr

Diffusing element	Saturation method	Saturation process†	Type of container	Saturation temperature, °C	Thickness of diffusion layer, mm	Phase composition of diffusion layer	Content of element on the surface, %	$H\mu$, daN/mm²	Surface quality, points, ‡
W	Solid	Contact	Nonhermetic	1200	0.06	α solid sol.	9.17	190	5
	"	"	Hermetic (in vacuo)		0.10	$\alpha + Fe_7W_6$	9.3	140	5
	Gas	Noncontact	Hermetic		0.050	$\alpha + Fe_7W_6$	15.8	210	6
Mo	Solid	Contact	Nonhermetic	1200	0.15	α solid sol. + oxides	8.24	200	4
	"	"	Hermetic (in vacuo)		0.17	α solid sol.	10.5	185	5
	Gas	Noncontact	Hermetic		0.075	$\alpha + Fe_7Mo_6$	18.2	280	6
V	Solid	Contact	Nonhermetic	1100	0.12	α solid sol.	8.17	255	2
	"	"	Hermetic (in vacuo)		0.12	α solid sol.	11.3	240	2
	Gas	Noncontact	Hermetic		0.14	α solid sol.	42.0	298	6
Cr	Vapor phase	Contact	Nonhermetic	1100	0.10	Cr_2N / α solid sol.	29.1(α)	193(α)	2
	"	Noncontact	"		0.025	α solid sol.	18.3	178	6
	Gas	Contact (in powder)	Contact		0.12	Cr_2N / α solid sol.	45.4(α)	1500/180(α)	6
	"	Noncontact	Hermetic		0.16	α solid sol.	74.0	195	6
	Liquid*	In molten salts	Crucible		0.15	α solid sol.	40–70	—	4
Al	Vapor phase	Gas	Nonhermetic	950	0.20	α solid sol.	8.17	770	1
	"	Noncontact	"		0.12	α solid sol.	6.8	715	5
	Gas	Contact (in powder)	"		0.15	α solid sol.	23.9	780	6

*Considered according to published data [16].

†Noncontact process of the gas method is carried out in a $H_2 + HCl$ atmosphere.

‡The highest point is denoted by 6, the lowest by 1.

Starting from this principle, the basic groups of diffusion saturation are defined as follows*: (1) Saturation from the solid phase (solid method); (2) saturation from the vapor phase (vapor-phase method); saturation from the gas phase (gas method); saturation from the liquid phase (liquid method).†

Solid Method. Saturation by the solid method takes place when the vapor pressure of the diffusing element is much less than the vapor pressure of the saturated metal. Solid pieces (or powders) of the diffusing substance come into contact with the surface of the part.

Saturation occurs by the contact process since transfer of the diffusing element to the surface is possible only through points of contact of the interacting metals.

The solid method is used for the saturation of iron, nickel and other metals by refractory metals: Molybdenum, tungsten, niobium, uranium, etc. The number of special cases of surface saturation of metals by the solid method includes chromizing or zinc-coating, carried out by heating a layer of these metals previously applied by some process or other (for example, electrolytically) to the surface of the base metal [2, 8, 14].

Vapor-Phase Method. Investigations have shown that in a number of cases, the transfer of the diffusion metal to the treated surface is produced by means of the vapor phase formed on heating the metal [7].‡ The vapor-phase method of saturation is carried out by contact and noncontact processes. In the first case, evaporation of the metal occurs in the reaction space in the immediate vicinity of the points of contact of pieces or powder of the evaporating metal with the surface of the treated metal; in the second case, evaporation of the metal in the reaction space occurs at some distance from the treated surface.§ Therefore, the vapor-phase method is used for the saturation of alloys by metals having a high vapor pressure compared with the treated alloy (for example, saturation of iron, nickel, molybdenum, and niobium by aluminum, chromium, magnanese, and zinc), and also by metals having a low vapor pressure (for example, saturation of iron and nickel by tungsten, niobium, and molybdenum).

For obtaining a higher surface quality of the treated part, it is advantageous to use the noncontact process, although the depth of saturation is then less (see Table 1). A vacuum is not always necessary.

The saturation of alloys by metals having a low vapor pressure requires the application of the noncontact process and is usually carried out under high-vacuum conditions, and also with separate heating of the treated metal and diffusing metal (the latter is heated to a much higher temperature than the saturated metal).

Data characterizing the vapor pressures of the different metals as a function of the evaporation temperature and permitting an assessment to be made of the possibility and conditions of carrying out the saturation of metals and alloys on their basis by different elements using the solid and vapor-phase methods are represented graphically (see Fig. 1).

---

* The bases of the classification of saturation methods were first laid down in [7].

† While recognizing the not altogether fortunate terminology of the saturation methods quoted in brackets, we shall make use of it in what follows, however, to avoid cumbersome discussions.

‡ Problems of the technology of the saturation methods discussed are not dealt with in the present article, since they are described in [1-23].

§ A vapor (in contrast to a gas) characterizes a substance in a state of staturation at a temperature below the critical temperature.

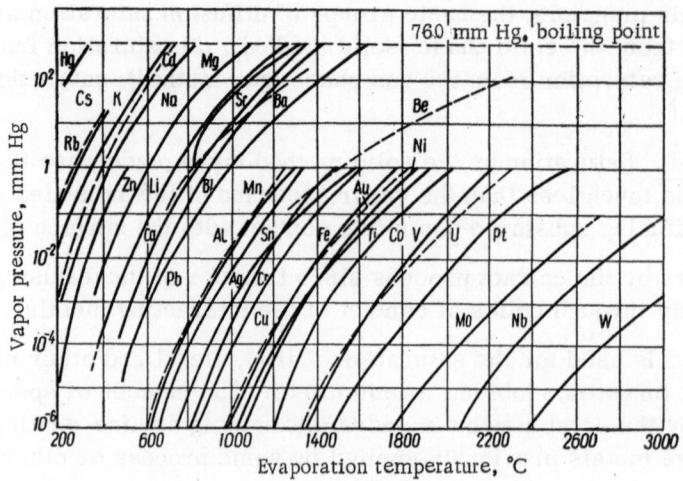

Fig. 1. Vapor pressure of various metals as a function of evaporated temperature.

Gas Method. The gas method of saturation is based on the interaction of the gas phase containing the diffusing element in the composition of a chemical compound, with the surface of the saturated metal. This interaction is accompanied by chemical reactions occurring at the metal – gas-phase interface and in the body of the gas phase. Various halides of the diffusing elements are used as active gas phase. The gas method may be carried out by contact (or pack process) or noncontact processes. In the first case, the gas phase is produced in the immediate vicinity of the surface of the treated part, as the result of the interaction of solid fractions of the metal in powder form with one of the halide gases (HCl, HF, HI, HBr, etc.) [1, 4, 6, 9, 10, 17, 23,], in the second case, the treated parts are situated in an atmosphere of the gas phase only, containing the halide of the diffusing metal [5, 19, 20].

In both cases, the following principal reactions may take place (for the case of chlorides):

Exchange reaction

$$MeCl_2 + M \rightleftarrows MCl_2 + Me, \tag{1}$$

Reducing reaction

$$MeCl_2 + H_2 \rightleftarrows 2HCl + Me, \tag{2}$$

Thermal decomposition reaction

$$MeCl_2 \rightleftarrows Me + Cl_2, \tag{3}$$

Disproportioning and nonproportioning reactions

$$2MeCl_2 \rightarrow Me + MeCl_4, \tag{4}$$

$$2MeCl_4 \rightarrow MeCl_3 + MeCl_5, \tag{5}$$

$$2MeCl_3 \rightarrow MeCl_2 + MeCl_4, \tag{6}$$

$$3MeCl_5 + 2Me \rightarrow 5MeCl_3, \tag{7}$$

where Me is the diffusing metal; M is the saturated metal.

In conducting saturation by the solid, vapor, or gas methods, the technical conditions involved play a considerable part. Attention should primarily be paid to the type of container. As shown in [4], the result of saturation by chromium depends essentially on the degree of hermetic sealing of the container.

If a nonhermetic container is used, not only is oxidation possible, but also saturation of the surface by nitrogen [4].

On the other hand, hermetic sealing of the container in saturation by the gas method (powder process) affects the kinetics of the process of saturation from the gas phase, since in this case there is no removal of the reaction products into the external atmosphere. This leads to a state of equilibrium in the gas medium as a result of which the concentration of the element on the surface and the depth of saturation are lower than with the use of a nonhermetic container under conditions when shifting of the reactions (1)-(3) to the right is quite probable. From this point of view, the noncontact gas process would appear to be the more effective, carried out in a hermetic container under conditions of continuous access of fresh portions of active gas and the removal of reaction products from the container (see Table 2).

Liquid Method. In the liquid method, saturation by the active phase, participating in the transfer of the diffusing element to the treated surface is either a melt of a salt containing the diffusing metal or a direct melt of the diffusing metal. In the first case, saturation is possible owing to a chemical reaction similar to reaction (1) at the metal—molten salt interface; in the second case, saturation occurs without chemical reaction directly from the molten metal. The liquid method is effected by immersion of the treated parts in a melt of salts of the diffusing metal [10, 11, 16] or in a melt of the metal [13, 22].

Table 2. Classification of Methods of Diffusion Saturation of the Surface of Alloys by Metals

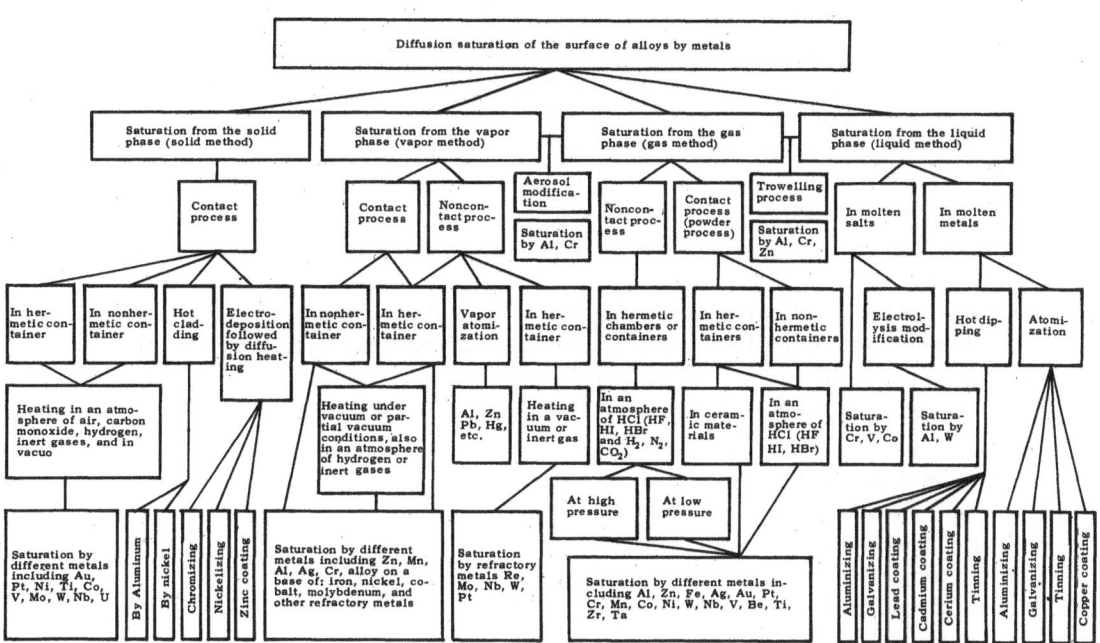

All the existing processes of saturation of metals by elements may be considered on the basis of the above-mentioned four methods.

The diagram opposite shows the calssification we have compiled of existing methods for the diffusion saturation of the surface of metals and alloys by various metals or metalloids. The diagram covers almost all the currently known processes and technical methods of saturating a surface by metals, and it may therefore be used both for practical purposes and for developing and analyzing new processes for the diffusion saturation of alloys by metals or metalloids.

Since the physicochemical characteristics of the active medium in saturation by metals using the aerosol process [12] or the trowelling process [21] are common to the vapor phase and gas methods, these methods of saturation are included in the diagram in the intermediate methods of saturation.

In conclusion, it should be remarked that the classification of methods of diffusion saturation of the surface of metals by elements makes it possible not only to systematize but also asses the possibilities of individual methods of saturation, and indicates the lines for the development of new advanced processes of diffusion saturation.

LITERATURE CITED

1. V. I. Archarov, Tr. Ural'sk Fil. Akad. Nauk SSSR Izd. Akad. Nauk SSSR, Moscow, No. 4 (1945).
2. V. I. Archarov and V. N. Konev, Vestn. Mashinostr., No. 11 (1955).
3. N. S. Gorbunov, Diffusion Coatings on Iron and Steel, Izd. Akad. Nauk SSSR, Moscow (1958).
4. Yu. N. Griboedov, Tr. Tsent. Nauchno-Issled. Inst. Tekhn. i Mash., No. 15 (1960).
5. G. N. Dubinin, Gas Method of Steel Chromizing, Metallurgizdat (1950).
6. G. N. Dubinin, Collection, "Research on Oxidation-Resistant Alloys," No. 138, Oborongiz, Moscow (1960).
7. G. N. Dubinin, Collection, "Science of Metals and Heat Treatment," Metallurgizdat, Moscow (1962).
8. N. A. Izgaryshev, Izv. Akad. Nauk SSSR Otd. Khim. Nauk, No. 6.(1941).
9. Yu. M. Lakhtin, Diffusion Metallization (Lecture No. 28 Correspondence Courses for the Improvement of Engineer Metallurgists and Heat-Treatment Engineers of VNITOMASh) Mashgiz, Moscow (1949).
10. A. N. Minkevich, Thermochemical Treatment of Steel, Mashgiz, Moscow (1950).
11. E. M. Morozova and E. D. Spivak, Heat Treatment and Machine-Tool Construction, Mashgiz, Moscow (1949).
12. V. I. Prosvirin, et al., Diffusion Metal-Carburizing in Aerosols, Inst. Tekhn.-Ékon. Inf. No. I, 56–110, (1956).
13. G. Rodoun, Protective Coatings with Metals Gostekhizdat, (1931).
14. P. Bardenheuer and R. Müller, Mitt. K. W. Inst. Eisenforsch., 14:295 (1932).
15. G. Becker et al, Stahl Eisen, No. 12 (1941).
16. J. Campbell et al, Trans. Electrochem. Soc., 96:262 (1949).
17. P. Galmiche, Rev. Mét., 3:193 (1050).
18. G. Gruble, Metallk. 19:438 (1950).
19. T. Hoar and E. Froom, J. Iron Steel Inst., 169:101 (1951).
20. F. Kelleg, Trans. Amer. Electrochem. Soc., 43:35 (1923).
21. N. Lockington, Metal Treat. Drop Forging, 24:136, 29 (1957).
22. W. Machu, Metallische Überzüge, Leipzig (1943).
23. R. Samuel and N. Lockington, Metal Treat., 18:70– 77 (1957).

THEORY OF DIFFUSION IN COMPOUNDS WITH A PREDOMINANTLY IONIC COMPONENT OF THE BOND FORCES

V. N. Konev

Systematic investigations of the influence of a second gas component on the mechanism and kinetics of reaction diffusion, and on the properties of the resulting reaction products provide the possibility, at a given stage, of separating from the numerous systems of "metal – two chemically active gases" type, those systems in which the reaction products are compounds with a predominantly ionic component of the bond forces. This group of systems has common relationships and a common physicochemical basis for the creation of a theoretical treatment of the experimentally observed relationships of the variation in properties of alloyed compounds produced under various thermodynamically equivalent conditions. The variation in properties of the reaction products is reflected in the diffusion characteristics of these compounds.

The present article is a first attempt to construct a model of a compound alloyed with another anion, and to examine the consequences following from this model.

The problem is complicated by the fact that scientific publications contain no data on the study of the variation in properties of compounds alloyed by an anion having another valence. This calls for the organization of special investigations.

In examining this question, we shall take as basis the hypotheses developed by Schottky, Wagner, and Hauffe [2] for nonstoichiometric compounds. Ionic and electronic disorder will be discussed, in agreement with chemical theory, from the corpuscular viewpoint, although this characterizes the given process quite restrictively. Indeed, although according to the equilibrium equations of the crystal lattice quoted below, it may also be assumed that the formation of the Me^{+++} ion occurs mainly by the detachment of an electron from an Me^{++} ion, this electron owing to the quantum mechanics tunnel effect ought to be determined according to many Me^{+++} ions. Since this also refers to the holes left in the Me^{++} ions after the detachment of the electron, i.e., to the formation of Me^{+++}, the character of the bonds is by no means the same as in the case of strict localization. Despite these inevitable limitations of the corpuscular model of disorder, the resulting consequences are still of considerable importance, and lead to results directly accessible to experimental verification. The results of experimental investigation are not only very important for the theoretical examination of electron disorder, but they are also of considerable applied value.

Figure 1 represents an example of a model of a nonstoichiometric compound with a deficiency of ions of the metal, according to Wagner. The difference in the proposed model is that not only the cations of this compound may be replaced by cations of another valence, but also the anions. It is particularly important to bear this in mind in the case of the formation of

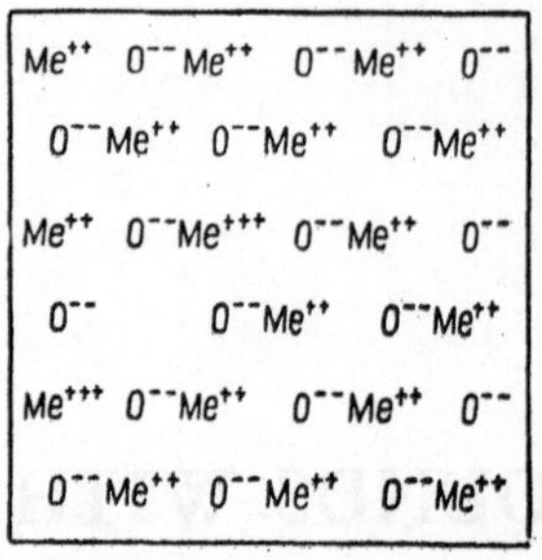

8   V. N. KONEV

Fig. 1. Model of disorder in ionic
lattice with metal deficiency.

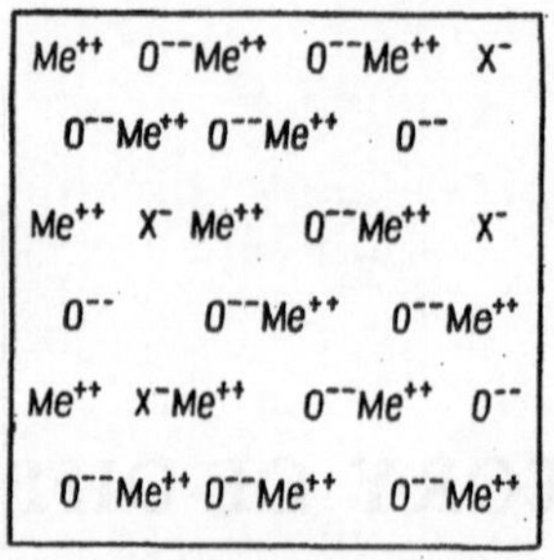

Fig. 2. Model of disorder of
the mixed phase Me (O, X).

metal oxides, into which may enter, under suitable equilibrium conditions, ions of sulfur, sele-
nium, and possibly even tellurium, i.e., for example, ions of the oxygen group of elements.
To simplify the representation of the scheme, we shall denote an ion of lower valence symbolically by X^-, although it must be noted that integral valence is by no means obligatory. For our
model, the important point is that the other anion has a lower valence. A sulfur anion, replacing
an oxygen ion in an oxide contributes less to the ionic component of the lattice bond forces than
does oxygen. The decrease in ionic component in compounds of metals with elements of the oxygen
group will occur from oxygen through to tellurium. It is only in this sense, in the scope of the corpuscular representation, that we shall understand what is meant by lower or higher valence of the anion.

It is very difficult to determine the true, exact value of the magnitude of the ionic component, and therefore the schematic representation in an approximation determined by the introduction of the above mentioned symbol (X^-) affects only quantitative calculations. With regard
to the qualitative side of the relationships, obtained as consequences of the proposed model, it
does not essentially undergo any variations through the introduction of such approximation.

As shown in Fig. 1, in the Wagner model, the cation lattice contains a certain number of
vacant sites, and electrical neutrality is established by the formation of cations of higher valence. The number of ions having a higher valence is equal to the number of electron defects,
while the number of the latter is determined by electron conduction (disregarding the conduction band).

If, in the crystal lattice of the above-mentioned compound, X^- ions are introduced instead
of O, then on the basis of a formal consideration of the possibility of electrical neutrality of the
crystal lattice, the following conclusions may be made.

1. The number of electron defects, i.e., the number of Me^{+++} cations, is diminished.
Consequently, electron conduction is diminished in comparison with the pure MeO. Since the
cation defects do not vary, the mobility of the metal ions practically also does not vary.

2. The number of cation defects increases, as shown symbolically in Fig. 2. Two X^- ions
create one Me^{++} vacancy and two electron "holes," on equilibrium being established, according
to the equation

$$(X)_2^{gas} + 2e^- \rightleftarrows \Box_{Me^{++}} + 2\Box_{e^-} + MeO_{(x_{sol})},\qquad(1)$$

where $\Box_{Me^{++}}$ are the vacant sites in the cation sublattice; $\Box_{e^-}$ are the electron holes.

If we ignore the other equilibrium perturbations of the lattice, the concentrations of ion
and electron defects under equilibrium conditions will always be in a constant relationship
$C\Box_{Me^{++}} = 2C\Box_{e^-}$. Assuming that the number of X^- ions injected into the lattice of the compound is proportional to the pressure P_{X_2}, the law of active masses ought to be satisfied. For
reaction (1) we then have

$$C\Box_{Me^{++}} \cdot C^2 \Box_{e^-} = KP_{X_2}.\qquad(2)$$

Taking into consideration the aforesaid constancy of the proportion of cation and electron defects, in the case of equilibrium with the gaseous atmosphere we get

$$C_\square = \text{const } P_{X_2}^{1/3}. \tag{3}$$

If we assume that conduction (cation and electron) $\mathcal{K}$ is directly proportional to the hole concentration, we have from (2) and (3)

$$\varkappa = \text{const } P_{X_2}^{1/3}. \tag{4}$$

If the cation component of conduction is due to the motion of cations in the lattice of the compound MeO, then for the increase in growth of the MeO layer on the metal in the presence in the oxidation medium of the gaseous component X_2, we get the analogous relationship

$$\frac{d\eta}{dt} = \text{const } P_{X_2}^{1/n}, \tag{5}$$

where $n = 3$ only if the anions X^- have a valence equal to 1, and if the conditions (1) and (2) are satisfied and the growth in the diffusion layer $MeO(X_{sol})$ occurs primarily as the result of the diffusion of cations in the crystal lattice.

The value of n as characteristic of the electron imperfection of the lattice of the compound $MeO(X_{sol})$ may be determined from the graph of the variation of the temperature coefficient of the thermo-emf with $\log P_X$, followed by calculation according to Hogarth's expression (1)

$$\frac{dE}{dT} = -\frac{k}{ne}\log P_{X_2}, \tag{6}$$

where k is the Boltzmann constant; e is the electron charge.

It should be noted that only the simultaneous application of several of the methods of investigating the properties of the alloyed compound MeO, (including chemical analysis, density determination, etc.) and the fact that these properties depend on certain parameters make it possible for us to ascertain the type of imperfections that arise when these compounds are alloyed with anions which have a different capacity for contributing to the ionic component of the bonding forces in the compound.

A similar examination is also possible for other nonstoichiometric compounds having other defects of the crystal lattice, for example for compounds having a metal excess in the interstices of the crystal lattice, and for compounds having vacancies in the anionic sublattice. Consequently, these model representations may be extended to all cases where under equilibrium conditions the crystal lattice is entered by anions having a different valence from the ions forming binary compounds of MeO type (i.e., alloyed with components having a lower (or higher) capacity for contributing to the ionic component of the bond forces of chemical compounds).

LITERATURE CITED

1. O. Kubaschewski and B. E. Hopkins, Oxidation of Metals and Alloys, Academic Press, New York (1953).
2. K. Hauffe, Reactions in Solids and on Their Surface [Russian Translation], IL, Moscow, Vol. I (1962), Vol. II (1963).

SOME CRYSTALLOGRAPHIC FEATURES OF
THE MECHANISM OF PHASE TRANSFORMATIONS
AND CHEMICAL REACTIONS IN SOLIDS

V. I. Arkharov

One of the principal questions forming part of the general problem of the mechanism of reaction diffusion in solids is the question of the mechanism of the formation of phase newly occurring in the reaction zone. The mechanism of the formation of the crystal nuclei of a new phase on the basis of the initial solid phase determines not only the orientation of the developing crystals of the new phase in relation to the crystals of the original phase, but also the fine structure (substructure) of the crystalline formations of the new phase. These structural factors, in their turn, affect many properties of the layers of the reaction product. Primarily, this is of importance for properties such as the mechanical properties (including the strength of the bond between the layer and substrate) and the diffusion properties. With regard to the latter, the function of the substructure is particularly important in a number of cases, due to the fact that diffusion along the boundaries of the structure elements or blocks is much easier than diffusion along sections of the more regular crystal lattice.

Collectively, these boundaries form a ramified network as it were of fine but wide channels of facilitated diffusion. The geometrical character of this network is of considerable importance for the overall diffusion effect. It determines the possibility of realizing the most direct paths for the diffusion flow which, while having general directivity, "spreads" along the network of the interblock boundaries. The geometrical character of this network is determined by the geometrical characteristics of the substructure blocks themselves and the character of their arrangement in the aggregate as a whole.

Apart from the absolute dimensions, particular significance is here assumed by the unequiaxial nature of the substructure elements, and the character of the regular orientation of these elements in relation to the substrate and to each other. This orientation of the block is connected in a complex manner with the crystallographic orientational relationships of the lattices of the original and newly formed phases.

As applied to the process of reaction diffusion, the mechanism of the formation of crystal nuclei and the formation of substructure aggregates from the primary crystal, forming in their totality a layer of the given phase, should be considered for two special processes:

1. When the atoms of the component of the external medium (X), in some way or other supplied to the external boundary (i.e., situated closer to the external medium) of the layer of the phase considered, enter the lattice of the phase, which after a limit concentration has been exceeded, are rearranged in the lattice of the new phase (of a higher compound). In particular,, this relates to the boundary of an oxide with a metal, when the penetration of the (X) atoms,

11

diffusing through the oxide, into the lattice of the metal, produce its rearrangement into the lattice of a lower oxide.

2. When the atoms of the metal, diffusing from the interior to the internal boundary (i.e., situated closer to the core) of the layer of the given phase, enter its lattice and bring about its rearrangement into the lattice of a less high oxide. This case includes the process, in which some of the (X) atoms, available in the composition of the lattice, leave it, and in the remaining zone of the initial phase, rearrangement of the lattice occurs with the formation of a new phase. Formally, the first case may be termed an oxidation reaction, and the second a reduction reaction in the solid phase.

In both these cases, the phase transformation includes a preparation stage and a rearrangement stage. In the first stage, changes occur in the composition of the atoms forming the lattice of the original phase; by diffusion into a given section of this lattice, additional atoms enter this section, or some of the complete set of atoms leave it. The second stage, in which rearrangement in the section of changed composition occurs nondiffusionally, combines the phenomenon of solid phase chemical reactions with ordered allotropic transformations in solids (of the type of the martensitic transformation). For the analysis of the mechanism of solid-phase chemical reactions, a detailed study of the process of the ordered rearrangement of the lattice is therefore of interest, special attention being paid to the formation of the block submicrostructure in this process.

In a number of articles [1-4], such a detailed study has been made by crystallographic calculations of the atomic displacements, resulting in the rearrangement of the lattice.

For allotropic ordered phase transformations, rearrangement of the lattice evidently occurs as follows:

First of all, in a small group of atoms, slight displacements occur by way of fluctuations, as a result of which, this group assumes a configuration, as if it had entered the lattice structure of a new phase.

Subsequently, displacements occur of the atoms closest to the rearranged group, these displacements being predetermined by the high energy stability of the resulting configuration. The original (incipient) region of the latter expands at the same time. Its further expansion occurs by "relay transfer," the displacement being transferred from atom to atom, starting from the original source, in each of the crystallographic directions in the crystal. The absolute value of the displacement increases in proportion to the distance from the original center of the rearrangement. The relay linkage of the atomic displacements leads to a large region of the original lattice being coherently rearranged into a large section of the lattice of the new phase.

The dimensions of such a region are determined by the break in coherence, the possibility of which increases with increase in the necessary displacement ξ, i.e., with movement from the rearrangement center. If ξ exceeds a certain critical value, there may be a disturbance of the regular relay transfer of the displacements to the following, more remote atoms, and here the coherent bond between the already rearranged part of the lattice and the part not yet rearranged will be broken.

It may be assumed that the critical value of ξ is of the order of the interatomic distance r_0 in the lattice.

Thus, from the fluctuation regrouping, commencing with the original center, a coherent rearrangement is developed in the limits of a certain region, on the periphery of which a rupture of coherence occurs and propagation of the rearrangement ceases, while in the interior, none of the atomic displacements exceeds r_0. Such a region, which to some extent is autonomous, we term a region of coherent lattice rearrangment; it is the primary crystal of the new phase.

Crystallographically, the external manifestation of the described rearrangement mechanism is the regular orientation of the crystal lattice of the new phase in relation to the original lattice. This orientational relationship is expressed by the following notation in crystallographic symbols:

$$(hkl)_{orig} || (h'k'l')_{new},$$
$$[uvw]_{orig} || [u'v'w']_{new}.$$

In the original lattice, as in the newly formed lattice, there are in the general case not one but several (a definite finite number) of systems of crystallographic phases of the same type, differing in their inclination to the axes of the coordinates (i.e., by the order of alternation and by the signs of the crystallographic indices in the symbol of the plane); the same applies to the crystallographic directions. Corresponding to one definite orientation of the original lattice, therefore, there are not one but several (a definite finite number) possible orientations of the lattice formed. All these possible orientations are crystallographically equivalent, but differ in their arrangement in space.

In each autonomous region of coherent lattice rearrangement, a definite orientation out of a number of possible ones is realized. Among the many autonomous regions of the new lattice, formed in the same crystallites of the original phase, individual regions may possess identical lattice orientations, and may be arranged parallel to each other; in the general case, however, they may differ in orientation (in the limits of the possible variants permitted by the general orientation law). In the latter case, the autonomous regions of coherent lattice rearrangement, as a rule, are arranged in relation to each other in directions which intersect each other in a complicated manner, forming a complex substructure in the limits of the rearranged part of the volume of one and the same crystallite of the original phase. Common to the structure of this part will be the fact that in it the crystal lattice is oriented according to one law, although also in its different variants realized in one or another elementary region. This structure may be described as a superposition of several textures, each of which is complete, all the textures coexisting in this aggregate of primary crystalline particles being related to each other by the crystallographic type.

The writer has developed a general method of calculating the dimensions and shape of the coherent lattice rearrangement [1]. The calculation is based on the magnitude δ_{100} of the difference in the periods of identity for the crystallographic directions $[uvw]_{orig}$ and $[u'v'w']_{new}$.

Assuming that at the center of the rearrangement there is an atom which is not displaced during the rearrangement we follow the displacement of atoms situated in the original lattice along the [uvw] direction passing through the stationary center. The two atoms closest to the center are displaced by $\pm\delta_{100}$, the next by $\pm n\delta_{100}$, where n is the number of the atom in the order of distance from the center. The condition for rupture of coherence $n\delta_{100} \leq r_0$ determines the length l of the section of the [uvw] row fitting into the limits of the region of coherent rearrangement.

Similarly, atoms are displaced in rows parallel to that considered (zero row), i.e., in rows referred to the same [uvw] direction of the same (zero) plane of the system $(khl)_{orig}$. However, the atom displaced the least, present in each row and corresponding to the central atom in zero row, is not situated in its row in the immediate vicinity of its center. Its position in each row of [uvw] direction may be determined by constructing the superposition of the crystallographic $(hkl)_{orig}$ and $(h'k'l')_{new}$ nets, while observing the second condition of the orientation bond

$$[uvw]_{orig} || [u'v'w']_{new}.$$

The determination of the position of an unknown atom, and also the calculation of the displacement of any atom may be done from formulas obtained by an examination of the superposition of the nets.

In view of the facts that in the rearrangement not only does the period of identity along the [uvw] now vary, but also the distance between adjacent rows in this direction, and the length of the sections of these rows fitting into the region of coherent rearrangement (in the zero plane) diminishes in the proportion in which each given row is distant from the zero row (passing through the center). By taking into consideration the relative displacements of the "centers" of each row and the indicated shortening of the sections, it is possible to construct a "coherence region" for the zero plane. A similar construction and calculation may also be made for any $(hkl)_{origin}$ plane of the same system, parallel to the zero plane, but in addition, taking into account the variation in the interplanar distances $d_{(hkl)orig} \neq d_{(h'k'l')new}$, we come to the conclusion that the regions of coherence in the different planes are different in size (they diminish as the plane becomes more remote from the zero plane), while the center of the region of coherence is not in the immediate vicinity of the zero center, but is displaced by a definite distance along the plane. This distance may also be calculated by examination of the superposition of the projections of the $(hkl)_{orig}$ and $(h'k'l')_{new}$ nets on the common (zero) plane.

The following data are required for the calculation: The crystallographic type of the original and newly formed lattice; the orientational relation observed between the lattices on rearrangement; the parameters (dimensional characteristics) of both lattices at the transformation temperature.*

On the basis of the general method, the regions of coherent rearrangement have been calculated for a number of specific cases. For example, in the case of the γ-α transformation in iron [2], the region of coherent rearrangement has the form of a very fine, long needle with an approximately equiaxial section, the width of which is 20-30 Å and the length 6500 Å.

For cobalt [3], the region of coherent rearrangement of the lattice by calculation is ribbon-shaped, with a thickness of 12, width about 1400 and length about 56,000 Å. In the α-β transformation in titanium [3], such a region has the respective dimensions 12 × 135 × 4900 Å.

The data quoted for the dimensions and form of the regions of coherent rearrangement are typical of transformations of corresponding crystallographic character: the first for the rearrangement of the face-centered cubic lattice into the body-centered cubic lattice, the second for the rearrangement of the body centered cubic lattice into the hexagonal close-packed lattice. Similar information concerning the regions of coherent rearrangement of the lattice in transformations, for which quantitative data necessary for the calculation of the dimensions and forms of the regions are available, are given in [4].

It is important to note that calculation has also shown the strong dependence of the dimensions of the regions of coherent rearrangement on the variation in chemical composition of the material. Thus, for the transformation of austenite into martensite in carbon steel [2], the length of the needle-shaped region varies from 6500 Å for iron to 95,000 Å for high-carbon steel (the cross-sectional dimensions of the needles vary very little).

It should also be noted that, as shown by a more detailed analysis, the regions of coherent rearrangement, regarded as primary elements of the submicrostructure of the material in

---

*It is noted that in cases where the constants of the lattices are known for another temperature (for example, room temperature), the coefficients of thermal expansion of both phases and the transformation temperature must be known for calculating the values of the constant at the transformation temperature.

different phase transformations, also differ with regard to their ability to form a regular relative arrangement in space.

In steel, these needle-shaped regions may form a crystallographically regular skeletal framework bounded by planes having a definite slope [5]. The "habitus" planes determine the boundaries of the "martensite needles," visible in the ordinary metallographic microscope and having a submicrostructure formed of the described skeletal framework, in which the interstices are overgrown with martensite formations, directly following the primary elementary region of coherent rearrangement.*

In cobalt, in view of the different crystallographic character of the lattice rearrangement, such linkage of the primary ribbon-like coherence regions cannot occur, and the pseudograin is formed by noncoherent layering of the autonomous regions.

As may be seen from the results of the calculation, the differences in shape and dimensions of the elements of the submicrostructure of martensite and phases similar to it may vary within wide limits, corresponding to the variation in composition of these phases and to the differences in crystallographic types of transformation. A still greater variety of characteristics of form and dimension is shown by the substructures of the products of solid-phase chemical reactions in view of the greater complexity of the crystallographic types of the lattices of the phases participating in such reactions, and the wide variation of composition in the process of these transformations.

The results of the analysis carried out for the more simple cases permit the assumption that further work in this direction may yield fresh useful information concerning reaction diffusion, especially in the processes of gaseous corrosion and also those of diffusion and thermochemical treatment of metals.

LITERATURE CITED

1. V. I. Arkharov, Fiz. Metal. i Metalloved. 12:853 (1961).
2. V. I. Arkharov, and Z. V. Korendyaseva, Fiz. Metal. i Metalloved. 14:5 (1962).
3. V. I. Arkharov and É. I. Kuznetsov, Izv. Akad. Nauk SSSR Otd. Tekhn. Nauk Met. i Toplivo No. 4 (1962).
4. V. I. Arkharov and É. I. Kuznetsov, Fiz. Metal. i Metalloved. 15:786 (1963).
5. V. I. Arkharov, Fiz. Metal. i Metalloved. 14:5 (1962).

---

*Thus, from our point of view, the martensite needle is not a single-crystal needle. Currently, there are experimental data in support of this point of view. It should be noted that the crystallographic indices which, according to calculation, should be ascribed to the habitus planes, agree with considerable accuracy with the experimental indices in the publications of a number of authors.

DETERMINATION OF DIFFUSION COEFFICIENTS AND LAYER THICKNESS OF THE PHASES IN REACTIVE DIFFUSION

V. T. Borisov, V. M. Golikov, and G. N. Dubinin

The saturation of the surface layers of a metal by various elements is often accompanied by the formation of several phases in the diffusion zone. The kinetics of the growths of the phases is determined by the reaction velocity (formation of oxides, carbides, intermetallic compounds, polymorphous transformation) and by the diffusion of atoms. It is usually assumed that in reactive diffusion, the process is limited by the diffusion of atoms [1, 4, 5].

In the present article, it is assumed that reactive diffusion occurs both at internal interfaces and on the external surface of the specimen. The article examines the various methods of determining the diffusion coefficients in the phases, and also the methods of estimating the thickness of the phase layers formed in diffusion saturation. Figure 1 is a diagram of a diffusion zone.

We assume that the concentrations C_k and C_k' at the phase boundaries in accordance with the phase diagram do not vary during phase growth, that the diffusion coefficient D_k in different phases does not depend on the concentration, and that the phase boundaries are displaced according to the law

$$y_k(t) = 2\xi_k \sqrt{D_k t}, \qquad y_k'(t) = 2\xi_k' \sqrt{D_k t}, \tag{1}$$

where ξ_k and ξ_k' are constant. The distribution of the diffusing substance inside the k-th is then given by the expression

$$C^k(x, t) = C_k - \delta_k \frac{\operatorname{erf} \dfrac{x}{2\sqrt{D_k t}} - \operatorname{erf} \xi_k}{\operatorname{erf} \xi_k' - \operatorname{erf} \xi_k}, \tag{2}$$

where

$$\delta_k = C_k - C_k', \quad \operatorname{erf} \xi = \frac{2}{\sqrt{\pi}} \int_0^\xi e^{-a^2} da.$$

It may be verified directly that in the region $y_k(t) \le \chi \le y_k'(t)$, this expression satisfies the diffusion equation $D C_{xx}^k = C_t^k$ and the necessary boundary conditions $C^k[y_k(t), t] = C_k$; $C^k[y_k(t), t] = C_k'$. For the first layer (k = 1), it is evident that $\xi \eta = 0$, and the quantity C_1 represents a concentration occurring on the external surface of the specimen. In contrast to other analogous constants, C_1 is not determined by the phase equilibrium diagram, but by the conditions of interaction of the specimen with the external medium. To be able to employ expression (2), therefore, it is necessary to observe in diffusion experiments conditions for which the concentration on the external surface of the specimen may be assumed with a sufficient degree of accuracy to be independent of time. The value of $k=n$ corresponds to the last layer, representing a solid solution of the element A in the base B. The extent of this layer is usually fairly large, and in the formulas, therefore, it should be assumed that $\xi'_n = \infty$, erf $\xi'_n = 1$. The constant quantity C'_n, representing the concentration of component A at $x = \infty$, still has the significance of the initial concentration of the diffusing element in the specimen. This follows from expression (2) for $t \to 0$: $C^k(x, 0) = C_n$; $0 \le x \le \infty$. Thus, expression (2) describes heterophase diffusion in the case where the concentration on all the phase boundaries and on the external surface of the specimen, the initial concentration and all the diffusion coefficients are constant quantities.

From relationships (2) and the expression for mass balance, we find the integral diffusion flow M through the section y situated in the k-th layer:

$$M = \frac{2\delta_k \sqrt{D_k t}}{\sqrt{\pi}\,(\text{erf}\,\xi'_k - \text{erf}\,\xi_k)}\, e^{-\frac{y^2}{4D_k t}}. \tag{3}$$

The ratio of the integral diffusion flows through the different sections y and y', situated in the limits of the same phase k (see Fig. 1) has the form $M/M' = \exp(\xi'^2 - \xi^2)$, and may be written as follows:

$$4D_k t = \frac{y'^2 - y^2}{\ln(M/M')}. \tag{4}$$

For the case where the quantities y, y_k, and y'_k and the product $(\delta_k D_k)$ are finite, but $\delta_k \to 0$ and $D_k \to \infty$, we get from relationship (3)

$$2(\delta_k D_k)t = M(y'_k - y_k). \tag{5}$$

The expressions obtained may be used for solving various types of problem. An important one is what is termed the direct problem: To determine the variation in concentration and the arrangement of all the phases, knowing all the diffusion coefficients, initial and surface concentrations, saturation time, and the phase diagram. Such a problem is presented, for example, in determining the optimum saturation diffusion conditions, when it is necessary to obtain predetermined thickness of phases on the surface of the part. According to (1) and (2), for this purpose it is sufficient to determine the quantities y_k and y'_k for all the zones. If, in the general formula (3), we put $y = y_k$ or $y = y'_k$, we obtain the integral flows M_k and M'_k through the front and back walls of the k-th layer (see Fig. 1). The figure shows that at each interface y_k, the relationship $M'_{k-1} - M_k = \Delta_k y_k$ $(\Delta_k = C'_{k-1} - C_k)$.

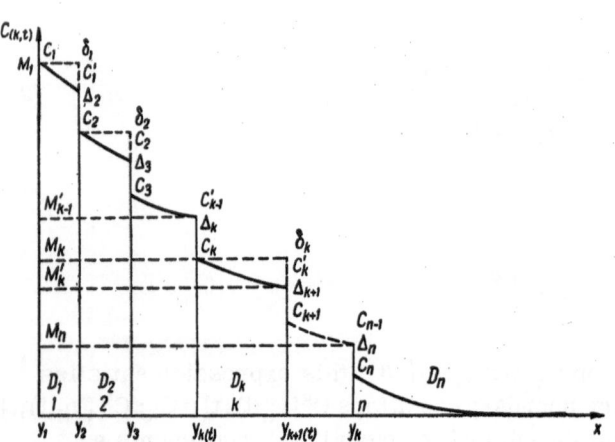

Fig. 1. Diagram of diffusion zone.

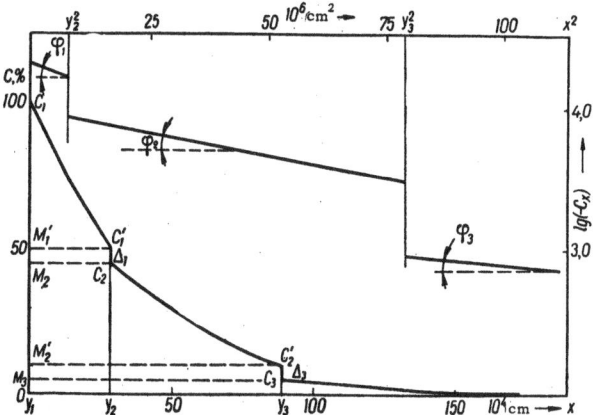

Fig. 2. Diffusion zone calculated on the basis
of known values of diffusion coefficients and
concentrations.

This gives the following system of equations (expressing the continuity of flow at the boundaries)
for finding the desired values of y_k:

$$\frac{2\delta_{k-1}\sqrt{D_{k-1}t}\,\exp\left(-\dfrac{y^2}{4D_{k-1}t}\right)}{\sqrt{\pi}\left(\operatorname{erf}\dfrac{y_k}{2\sqrt{D_{k-1}t}}-\operatorname{erf}\dfrac{y_{k-1}}{2\sqrt{D_{k-1}t}}\right)}-\frac{2\delta_k\sqrt{D_kt}\,\exp\left(-\dfrac{y_k}{4D_kt}\right)}{\sqrt{\pi}\left(\operatorname{erf}\dfrac{y_{k+1}}{2\sqrt{D_kt}}-\operatorname{erf}\dfrac{y_k}{2\sqrt{D_kt}}\right)}=\Delta_k y_k. \tag{6}$$

Here $k=2, 3,\dots, n$; $y_1=0$; $y_{n+1}=\infty$; furthermore, the obvious equality $y_k'=y_{k+1}'$ (see Fig. 1) is
taken into account.

For a two-layer or three-layer medium, if the method of successive elimination is used,
this system is solved numerically comparatively simply; with increase in n, the solutions become rather cumbersome. By way of illustration, Fig. 2 shows the results of the calculation of
the distribution of the component A and the position of the phase boundaries. In performing the
calculations, the following assumptions were made: $C_1=100$, $C_1^1=50$, $C_2=45$, $C_2^1=10$, $C_3=5\%$
(these values should be taken from the phase diagram), $t=25$ hr, and the following values of the
diffusion coefficients (should be known beforehand): $D_1=10^{-10}$, $D_2=2\times10^{-10}$, and $D_3=4\times10^{-10}$
cm^2/sec.

In accordance with the accepted notation, $\delta_1=50$, $\delta_2=35$, $\delta_3=5\%$; $\Delta_2=5$, $\Delta_3=5\%$. The following was obtained as a result of the solution of equation (6): $y_2=28\times10^{-4}$; $y_3=89\times10^{-4}$ cm,
which makes it possible by the use of formulas (1) to calculate all the values, and in accordance
with expression (2) to write the expression for the concentration in the zones $C^1=100-200$ erf
167 x, $C^2=69.6-69.2$ erf 118 x, $C^3=17.0$ erf 83 x.

As will be seen from Fig. 2 (for y_2 and y_3), the thicknesses of the phase layers are:
$h_1=y_2=28\times10^{-4}$ cm; $h_2=y_3-y_2=61\times10^{-4}$ cm.

It is of interest to examine the system (6) in linear approximation, when all the quantities
of the form $y/2\sqrt{Dt}$, except $y_{n+1}=\infty$, are small. In this case, in all the layers $k\neq n$, as will be
seen from formula (2), the variation in concentration is represented by a straight line. If, in
the transformation of the system, we confirm ourselves always to terms of 1st order of small-

ness, the system assumes the form

$$y_{k+1} - y_k = \frac{\delta_k D_k}{\delta_{k-1} D_{k-1}} (y_k - y_{k-1}) \quad k = 2, 3 \ldots (n-1),$$

$$\sqrt{\pi D_n t} - y_n = \frac{\delta_n D_n}{\delta_{n-1} D_{n-1}} (y_n - y_{n-1}) \quad k = n \tag{7}$$

and has the following solution

$$y_k = \frac{\sum_{i=1}^{k-1} \delta_i D_i}{\sum_{i=1}^{n} \delta_i D_i} \sqrt{\pi D_n t} \quad \text{or} \quad h_k = \frac{\sqrt{\pi D_n t}}{\sum_{i=0}^{n} \delta_i D_i} \delta_k D_k \, k = 1, 2 \ldots (n-1), \tag{8}$$

where $h_k = y_{k+1} - y_k$ represents the thickness of the k-th layer. The expression obtained is approximate, and in cases in which it is insufficient, the accuracy may be increased by solving the system (6). In doing this, it must be borne in mind that the value of the approximate solution (8) substantially facilitates calculation of the quantities y_k. The approximation (7) corresponds to steady-state diffusion, examined in [1].

In some cases, for the determination of diffusion coefficients in the phases, it is assumed that the dependence of the thickness of the k-th layer on time may be represented by the expression $h_k = \gamma_k \sqrt{D_k t}$, where $\gamma_k \approx 1$ [5]. Formula (8) shows that at least in steady-state approximation, the layer thickness is proportional to the first power of the diffusion coefficient in the given layer and lies between zero and the quantity $\sqrt{\pi D_n t}$, where D_n is the diffusion coefficient in the last zone (solid solution). The above mentioned method, therefore, can evidently merely provide a general estimate of the magnitudes of the diffusion coefficients in the entire system as a whole, and is not suitable for determining the quantitative relationships between the diffusion coefficients in the different phases.

Intermediate phases, for example stoichiometric chemical compounds, in many cases have very low solubility of the components. If such a compound is situated in the layer m, the quantity δ_m is quite small and in the experimental determination of the variation of concentration in such a layer, an almost horizontal straight line is obtained.

The diffusion flow in the layer D_m, $C_x^m \approx \delta_m D_m$, has a finite value if the extent of such a layer is finite. Therefore, a low value of δ_m ought to be compensated correspondingly by high values of the diffusion coefficient D_m. For $\delta_m \to 0$, $D_m \to \infty$ ($\delta_m D_m$ is limited), the expressions for the integral flows have the form of expression (5), while the corresponding terms in the system (6), containing the diffusion coefficient D_m of the given layer, assume the form $2(\delta_m D_m) t / (y_{m+1} - y_m)$. The said expression contains only the combination $\delta_m D_m$. If this quantity is known, the values of the flows and the displacement of the boundaries of the corresponding layer are determined completely [after solving the system (6)]. The product $\delta_m D_m$ is therefore used as quantity characterizing the diffusion properties of a stoichiometric (or almost stoichiometric) compound [3].

In some cases, the direct problem formulated in the foregoing may be put more definitely. For example, it is possible to raise the question concerning the search for values of the temperature T and duration of diffusion saturation for predetermined layer thickness of the phases formed. Such a postulation has meaning only when it is a question of determining the thicknesses of only two phases, since there are only two free parameters T and t. The equations quoted enable the desired values of T and t to be found. For example, let the quantities h_1 and

Table 1

x	$\sqrt{\pi}\,xe^{x^2}\,\mathrm{erf}\,cx$	x	$\sqrt{\pi}\,xe^{x^2}\,\mathrm{erf}\,cx$
0.1	0.159	1.2	0.804
0.2	0.286	1.4	0.839
0.3	0.390	1.6	0.866
0.4	0.475	1.8	0.887
0.5	0.545	2.0	0.904
0.6	0.603	2.2	0.917
0.7	0.652	2.4	0.929
0.8	0.693	2.6	0.936
0.9	0.720	2.8	0.944
1.0	0.757	3.0	0.950

h_2 be known. Taking into account the dependency of the diffusion coefficient D_k and the quantity δ_k on the temperature, we may write making use, for example, of expression (8):

$$h_1 = \sqrt{t}\,\varphi(T)\,\delta_1 D_1(T) \ \text{ and } \ h_2 = \sqrt{t}\,\varphi(T)\,\delta_2 D_2(T).$$

The temperature is determined from the ratio

$$h_1/h_2 = \delta_1 D_1(T)/\delta_2 D_2(T),$$

and the time from either of the two equations.

The inverse problem with respect to that formulated in the foregoing is to find the diffusion coefficients in all the phases from such characteristics of the heterophase diffusion process as may be determined experimentally. This problem is important not only with regard to the development of methods of determining diffusion coefficients and the possibility of collecting information regarding the value of the diffusion parameters in different alloys. A knowledge of diffusion coefficients is necessary for solving the problems discussed in the foregoing connected with the search for the optimum conditions for producing diffusion coatings or the necessary arrangement of the phases. It should be noted that the diffusion coefficients contained in the quoted formulas, in particular for high concentrations of the diffusing element, characterize not only its mobility in the basic lattice of the phase, but also take into account, in a definite manner, the diffusion of the atoms of the basic solvent [4], and may differ in one degree or another from the values obtained by other methods. The possibility of determining the diffusion characteristics δD of the phases having quite low solubility of the components (compounds close to the stoichiometric), is also of value. The determination of these characteristics in the use of other diffusion methods may be found difficult (in view of the low solubility).

In what follows, various methods are given for the determination of diffusion coefficients for known concentration distribution of the diffusing substance in all the layers of the surface zone.

1. If the distribution of the substance is known and graphical differentiation may be performed ($\delta \neq 0$ and nonsteady state of the diffusion conditions), the diffusion coefficients may be determined by the use of relationship (2):

$$\ln C_x^k(x,t) = -\frac{x^2}{4D_k t} + \text{const}, \quad 4D_k t = \frac{1}{\tan \varphi}. \tag{9}$$

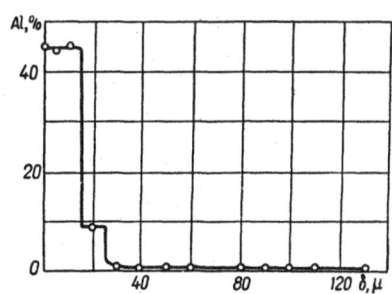

Fig. 3. Distribution of aluminum in molybdenum.

The graph of the variation of the function $\ln C_x^k$ with the argument x^2 consists of sections of straight lines, each of which corresponds to a definite phase, and the angle φ of slope of a straight line is connected with the diffusion coefficient by the usual relationship. In Fig. 2, this method of treatment is illustrated by the construction of the function (9) in the form straight-line sections for each phase of the diffusion zone.

2. In the case where the concentration distrubution permits determination of the integral

diffusion flows M_k and M_k', representing areas of corresponding parts of the diffusion zone (see Fig. 1), the diffusion coefficient in the k-th layer may be found from the formula

$$4D_k t = \frac{y_{k+1}^2 - y_k^2}{\ln(M_k/M_k')}, \tag{10}$$

this being a special case of the relationship (4) for $y' \to y_k' = y_{k+1}, y \to y_k$. This variant is examined in greater detail in an article by the present author, in which it is shown that formula (10) is valid for all the layers except the last. For $k = n$, the calculation is performed otherwise:

$$2\xi_n \sqrt{D_n t} = y_n, \tag{11}$$

where

$$\sqrt{\pi} \xi_n e^{\xi_n^2} \operatorname{erf} C\xi_n = \frac{C_n y_n}{M_n}.$$

The function $F(\xi) = \sqrt{\pi} \xi_n e^{\xi_n^2} \operatorname{erf} C\xi_n$ may be tabulated or represented graphically. The results of tabulation are given in Table 1.

Knowing C_n, y_n, and M_n from experiment, it is possible to find the value of the quantity ξ_n, and then also of D_n, the diffusion coefficient in the last layer of the diffusion zone. It should be noted that when using this method of determining diffusion coefficients, it is possible to take arbitrary sections y, y' situated in one phase [4]; it is not necessary to take them across the phase boundary.

As an illustration of this method of treatment, let us examine the graph (see Fig. 2) from which we find: $M_1 = 0.381$ (area of figure $OC_1C_1'C_2C_2'C_3C_3'O$), $M_1' = 0.315$ (area $OM_1'C_1'C_2C_2'C_3C_3'O$), $M_2 = 0.301$ (area $OM_2C_2C_2'C_3C_3'O$), $M_2' = 0.113$ (area $OM_2'C_2'C_3C_3'O$), $M_3 = 0.067\%$ cm (area $OM_3C_3C_3'O$); $C_3y_3 = 0.045\%$ cm; coordinates of the phase boundaries $y_2 = 28 \cdot 10^{-4}$, $y_3 = 89 \cdot 10^{-4}$ cm.

Calculations for D_1 and D_2 (for $t = 25$ hr) performed according to formula (10) give $1 \cdot 1 \times 10^{-10}$ and $2 \cdot 1 \times 10^{-10}$ cm^2/sec, respectively. The diffusion coefficient D_3 in the last zone is determined from expression (11) $C_3y_3/M_3 = 0.662$, from this value of the function we find the value of the argument $\xi_3 = 0.723$, which gives $D_3 = 4.2 \times 10^{-10}$ cm^2/sec.

3. For phases having very low solubility of the components, when the concentration of the diffusing element practically does not vary over the extent of the layer, the diffusion parameter $\delta_k D_k$ may be determined from the relationship (5):

$$2(\delta_k D_k) t = M_k(y_{k+1} - y_k).$$

As illustration of the method of calculating this characteristic, let us examine the aluminizing of molybdenum from the gas phase at 1100°C for 6 hr by the method described in [2]. The distribution of aluminum in the diffusion saturation of molybdenum is shown in Fig. 3. From the figure, we have: $C_1 = 45$; $C_1' = 45$; $C_2 = 9$; $C_2' = 9$; $C_3 = 1.5\%$; $\delta_1 = 0$, $\delta_2 = 0$, $\delta_3 = 1.5\%$; $\Delta_2 = 36$, $\Delta_3 = 7.5\%$; $y_1 = 0$, $y_2 = 15 \times 10^{-4}$, $y_3 = 25 \times 10^{-4}$ cm; $M_1 = 807 \times 10^{-4}$, $M_2 = 225 \times 10^{-4}$, $M_3 = 80 \times 10^{-4}\%$ · cm; $C_3y_3 = 38 \times 10^{-4}\%$ · cm.

Utilizing these data, we find from formula (5):

$$(\delta_1 D_1) = \frac{M_1(y_2 - y_1)}{2t} = 2.8 \cdot 10^{-9} \frac{\% \cdot \text{cm}^2}{\text{sec}},$$

$$(\delta_2 D_2) = \frac{M_2(y_3 - y_2)}{2t} = 6.2 \cdot 10^{-10} \frac{\% \cdot \text{cm}^2}{\text{sec}}.$$

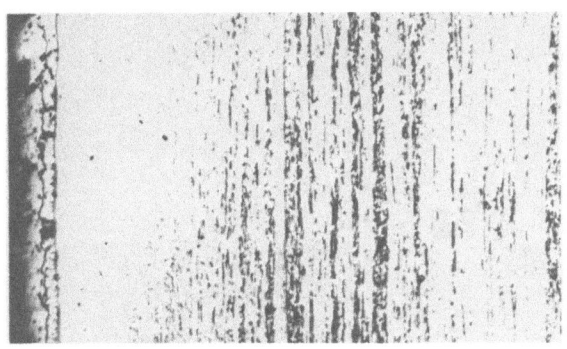

Fig. 4. Microstructure of diffusion zone of molybdenum after aluminizing.

For determining the coefficient of diffusion in the last zone, we make use of formula (11) as before, $C_3 y_3/M_3 = 0.472$ and find $\xi = 0.4$, which gives the value of $D_3 = 4.5 \times 10^{-10}$ cm^2/sec for the diffusion coefficients of aluminum in molybdenum.

Figure 4 shows the microstructure of the diffusion zone, enabling the position of the phase boundaries to be determined, the crystal structure of the first two phases being unknown.

4. The variants examined in the foregoing did not require in principle a knowledge of the phase diagram of the system examined. If, however, all the concentrations on the phase boundaries are known, for determining the diffusion coefficients on all the phases (or the parameters $\delta_k D_k$ for compounds of stoichiometric compostion) it is sufficient to know the thicknesses of the phase layers and the total area of the diffusion zone, or the value of variation in weight of the specimen, which is proportional to that area. By using radioactive isotopes, the total mass may be determined from the total γ-radioactivity of the specimen.

The expression for the total mass obtained from (3) for $y = y_1 = 0$,

$$M_1 = 2\delta_1 \sqrt{D_1 t}/\sqrt{\pi}\, \mathrm{erf}\,(y_2/2\sqrt{D_1 t}),$$

permits the value of D_1 to be found directly.

The subsequent coefficients D_2, D_3......, D_n may then be obtained by means of the recurrent equation (6), connecting D_k directly with the value of D_{k-1}. It will also be noted that by means of equation (6), it is possible to find from any one diffusion coefficient all the others, if the phase diagram is known and the thicknesses of the phase layers are determined.

In using the fourth variant for the treatment of the curve shown in Fig. 3, we determine $\delta_1 D_1$ from the value of M. It is noted that under our conditions, formulas (5) and (12) do not differ, since δ is small and δD is limited. For the same reason, equation (6) has the form (k = 2, 3)

$$\frac{2(\delta_1 D_1)t}{y_2} - \frac{2(\delta_2 D_2)t}{y_3 - y_2} = \Delta_2 y_2;$$

$$\frac{2(\delta_2 D_2)t}{y_3 - y_2} - \frac{2\delta_3 \sqrt{d_3 t}\, e^{-y_3^2/4D_3 t}}{\sqrt{\pi}\, \mathrm{erf}\, C^{y_3/2\sqrt{D_3 t}}} = \Delta_3 y_3.$$

(13)

Substituting the values of $(\delta_1 D_1)$ in the first equation, we find $\delta_2 D_2 = 6.2 \times 10^{-10}\% \cdot$ cm^2/sec, and from the second equation $D_3 = 4.4 \times 10^{-10}$ cm^2/sec, i.e., variants (3) and (4) give similar values.

LITERATURE CITED

1.   V. I. Arkharov, Fiz. Metal. i Metalloved. 8:193 (1959).
2.   G. N. Dubinin, Dokl. Akad. Nauk SSSR 84:5 (1952).
3.   G. S. Kreimer, L. D. Éfros, and E. A. Voronkova, Zh. Tekhn. Fiz. XXII:859 (1952)
4.   N. M. Rodigin, Fiz. Metal. i Metalloved. 11:240 (1961).
5.   H. Bückle, Rech. Aeron. 16:61 (1950).

GLOW-DISCHARGE SILICONIZING OF METALS

D. A. Prokoshkin, B. N. Arzamasov,
and E. V. Ryabchenko

In many cases siliconizing provides a reliable coating for the protection of metals from gaseous or chemical corrosion. In Particular, siliconizing is one of the methods of protecting regractory metals, such as molybdenum, tungsten, niobium and tantalum, from high-temperature oxidation.

The known methods of siliconizing have substantial drawbacks. They require high-temperature furnaces for carrying out the processes, and are also accompanied by a reduction in strength of the metals, since the necessary coatings are produced at temperatures above the recrystallization temperatures.

We have developed a new process of siliconizing metals (in the glow-discharge), which is free from the above-mentioned drawbacks, and differs in principle from the known processes with regard to both the heating method and the physical phenomena accompanying the siliconizing process.

The process of glow-discharge heating of metals has only recently found practical application, having been used in particular for the nitriding and carburizing of steel [2-5, 7, 8].

Published references [4, 5, 7] indicate that glow-discharge nitriding and carburizing under certain conditions proceed more rapidly than in the known saturation processes at the corresponding temperature. This acceleration of the process is due to the presence of an ionized gaseous atmosphere [5, 7].

The present article describes the results of experiments carried out on glow-discharge siliconizing of refractory metals. Glow-discharge siliconizing was carried out by the uniflow gas method in an atmosphere consisting of silicon tetrachloride vapor and hydrogen. Rectified silicon tetrachloride and commercially pure hydrogen without additional purification were used.

Molybdenum, tungsten, niobium and tantalum were subjected to siliconizing. All the preparatory experiments and development of the working conditions were carried out on Mark MRN molybdenum. The specimens were prepared in the form of a cylinder 6 mm in height and diameter, with a bottomed hole 2.5 mm in diameter drilled in the end face for the insertion of a thermocouple.

Figure 1 shows diagrammatically the experimental apparatus used for glow-discharge coating. The principal part of the apparatus is the reaction chamber 1, made of molybdenum glass. At the top is a fused-in anode 2. At the bottom, a ground-in stopper 3, comprising fused-in leads for the cathode and thermocouple, is inserted. The specimen 4 is placed on a molybdenum cathode 5 serving as support. The distance between the anode and cathode is 12-16 mm. The molybdenum rod 6 of the cathode is insulated from the chamber atmosphere by a porcelain tube. The platinum/platinum-rhodium thermocouple 7 is insulated throughout its entire length

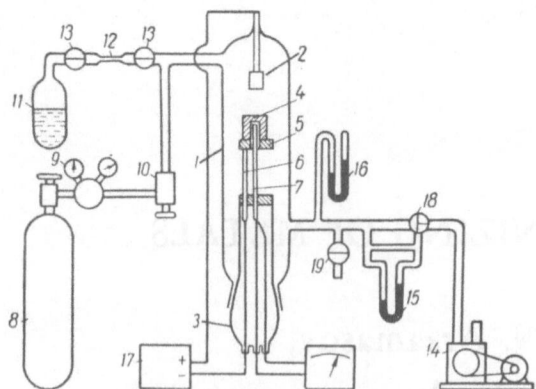

Fig. 1. Diagram of apparatus used the glow-discharge siliconizing of metals.

by a quartz sheath. Insulation is necessary so that the cathode part of the glow-discharge comes into contact only with the specimen and support.

The hydrogen is supplied to the discharge chamber from a cylinder 8 through a reducing valve 9 and needle valve 10. The silicon vapor, produced by evaporation from the vessel 11, enters the chamber through the capillary tube 12 and vacuum cock 13.

The gases are exhausted from the system by the rotary pump 14. The flow rate of the gases is measured by means of the differential manometer 15, and the pressure by means of the mercury manometer 16.

The 5 kW constant voltage source 17 ensures full-wave rectification without smoothing.

After the specimen to be coated has been placed in position, the air is evacuated from the system to a pressure of 1-0.1 mm Hg. The system is then flushed with hydrogen, which is passed through the system continuously at a pressure of 20 mm Hg, the specimen being heated at this pressure by the glow-discharge to a temperature of 1000°C in the course of 1 min. This results in the preparatory removal from the specimen of oxide films and incidental contamination by cathode sputtering and reduction by hydrogen. The necessary flow rate and pressure of the hydrogen and silicon tetrachloride are adjusted by means of the needle valve, cock 18 and cocks 13. When dynamic equilibrium of the gases has been reached, the glow-discharge is switched on. The glow discharge burns at 500-700 V. Its color is violet. The specimen and support (i.e., the cathode), subjected to bombardment by ions of the gases, are rapidly heated to the necessary temperature. Rapid and uniform heating of the specimen throughout its entire cross section is ensured by reason of its small dimensions and the sufficiently high thermal conductivity of the molybdenum.

It should be noted that the glow discharge does not heat the other parts of the reaction chamber. The chamber is heated only insignificantly by radiation from the specimen. For attaining temperatures of 1000-1200°C, a comparatively low current density of the order of 100-150 mA/cm² is required. After heating for a definite time and cooling the specimen, the gases are evacuated from the chamber, and air is allowed to enter the system through the cock 19.

Preliminary experiments showed that the rate of diffusion saturation of metals by silicon in the glow-discharge depended to a considerable degree on the pressure in the chamber, the ratio by volume of silicon tetrachloride vapor and hydrogen in the mixture, their flow rate and the temperature of the saturated specimen. In the majority of the experiments, the pressure was 40 mm Hg and the flow rate of the gases not more than 0.5 liter/min.

It was found that under optimum conditions, the rate of glow-discharge siliconizing of metals exceeded the rate of ordinary gas siliconizing in an atmosphere of silicon tetrachloride and hydrogen.

Figure 2 shows the variation of coating thickness with duration of glow-discharge siliconizing at 1000°C (3), and for comparison in a similar atmosphere but by the ordinary uniflow gas method (1), and circulation gas method (2). As will be seen from the graph, the rate of glow-discharge siliconizing considerably exceeds that of ordinary gas siliconizing. In the glow-discharge siliconizing of molybdenum, a coating 25 μ thick is formed in 5 min, and one

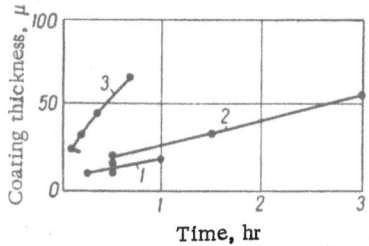

Fig. 2. Variation of coating thickness with duration of molybdenum siliconizing in an atmosphere of $SiCl_4$ and H_2 at 1000°C.

67μ thick in 40 min. A high rate of glow-discharge siliconizing is also found in the saturation of tungsten, niobium, and tantalum. Investigations of the glow-discharge siliconizing conditions are being continued.

Microscopic and x-ray structural analysis studies as well as microhardness measurements showed that in the glow-discharge siliconizing of molybdenum, tungsten, niobium, and tantalum, the coating mainly consisted of disilicides of the respective metals. In addition to an outer silicide layer, inner, intermediate layers of a silicide coating with a lower silicon content were found.

Figure 3 shows the microstructure of tungsten (b), niobium (c) and tantalum (d), glow-discharge siliconized at 1000°C for 30 min, and the same for molybdenum (a) for 40 min. Three diffusion layers are clearly visible on the molybdenum. The outer, thickest layer of $MoSi_2$ has a coarse-grained columnar structure, the middle layer is very thin, and the inner layer has appreciable thickness. Two diffusion layers are observed on niobium and tantalum. On niobium, the outer layer of $NbSi_2$ has a columnar structure, the inner layer is a thin. On tantalum, the outer layer of $TaSi_2$ has a coarse-grained columnar structure. On tungsten, only one layer of WSi_2 is to be seen. The following etchants were used for the coatings; For Mo and W, 95% HF, 5% HNO_3; 10% $K_3Fe(CN)_6$, 10% KOH, 80% H_2O; for Nb and Ta, 50% HF, 50% HNO_3.

X-ray structural analysis studies showed that the disilicides formed had a tetragonal lattice in the case of siliconizing of molybdenum and tungsten, and a hexagonal lattice in the case of niobium and tantalum. Figure 4 shows the x-ray patterns of $MoSi_2$ (a), WSi_2 (b), $NbSi_2$ (c) and $TaSi_2$ (d), taken in chromium radiation. The x-ray patterns obtained were in full agreement with the theoretical, with regard to both the position and intensity of the lines.

The microhardness measurements of the diffusion layers under a load of 50g showed the following results. Hardness of $MoSi_2$, 1565; inner layer next to the molybdenum, 1350; WSi_2, 1865; and $NbSi_2$, 1125 daN/mm^2.

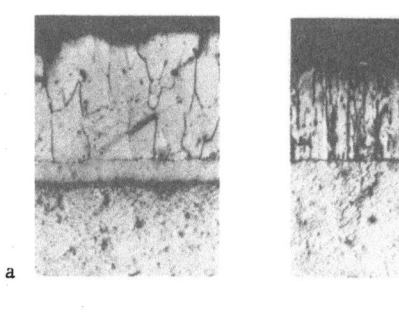

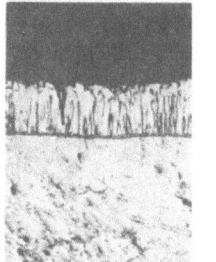

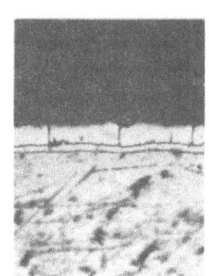

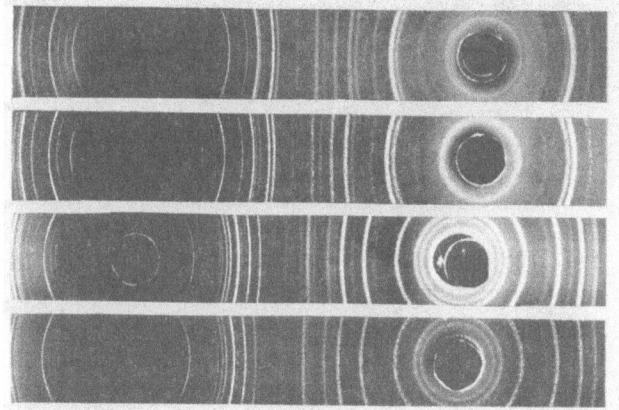

Fig. 3. Microstructure of refractory metals glow-discharge siliconized at 1000°C (× 340)

Fig. 4. X-ray patterns of the upper layers of coatings on glow-discharge siliconized refractory metals. Chromium radiation without filter.

SUMMARY

It has been found possible to saturate metals with silicon in the glow-discharge. In the glow-discharge siliconizing of refractory metals at a temperature of 1000°C, the outer layer of the coating consists of disilicide of the metal, below which are intermediate layers with a lower silicon content. The rate of glow-discharge siliconizing of metals considerably exceeds the rate of ordinary gas siliconizing.

Glow-discharge siliconizing of metals may be conducted at temperatures below the recrystallization temperature, enabling the part to be kept in the hardened state. The high activity of the glow-discharge siliconizing process may be explained as being due to the presence of an ionized atmosphere and to the activated condition of the surface of the coated metal.

The process is economical. It does not require high-temperature furnaces, since only the coated part is heated by the glow discharge.

LITERATURE CITED

1.   B. N. Arzamasov, and D. A. Prokoshkin, "New and Improved Processes of Heat and Chemical-Heat Treatment," Advanced Scientific, Technical, and Industrial Experience, 7-63-143/4, GOSINTI, (1963).
2.   V. S. Vanin, Izv. Akad. Nauk SSSR. Otd. Tekhn. Nauk Met. i Toplivo, No. 3 (1960).
3.   V. S. Vanin, Metalloved. i Term. Obrabotka Metal., No. 8 (1961).
4.   V. S. Vanin, Izv. Akad. Nauk SSSR. Otd. Met. Tekhn. Nauk Toplivo, No. 5 (1962).
5.   F. Barbas, L'Usine Nouvelle, special issue (1960).
6.   E. Fitzer, Berg- und Hüttenmänn. Manatsh. Montan. Hochschule Leoben 97:581 (1962).
7.   H. Knuppel, Stahl Eisen 78(26):1871 (1958).
8.   T. Noren and L. Kindbom, Stahl Eisen 78(26):1881 (1958).

VACUUM SILICONIZING OF REFRACTORY METALS

V. E. Ivanov, E. P. Nechiporenko, V. I. Zmii, and V. M. Krivoruchko

Several methods have been described for the siliconizing of metals [6-9]. Most of them are associated with the presence of $SiCl_4 + H_2$, $Si + NH_4Cl$, or melts of $Cu + Si$, $Al + Si$.

One of the methods of the vacuum siliconizing of refractory metals, the thermodiffusion saturation of a metal with silicon at a residual pressure of $1 \cdot 10^{-5}$ mm Hg, is described in [4].

The object of the present work is to study the kinetics and mechanism (the function of the vapor phase, velocity of the chemical reaction on the surface, diffusion, value of the solid contact, influence of temperature gradient between the cell and specimen) of the formation of thermodiffusion layers on refractory metals in vacuum siliconizing.

The diffusion study was conducted on the Mo-Si system. The specimens were molybdenum tablets $40 \times 40 \times 1$ mm in size. Silicon of 99.99% purity in powder form, particle size $5-7\mu$, was used for the experiments. The specimens were placed in a molybdenum dish packed tightly in the powder. The dish was charged through a forechamber into the furnace heated to the required temperature by a molybdenum heater. The experiments were carried out at temperatures of 1200-1350°C and at a residual pressure of 1.10^{-5} mm Hg. The diffusion layer on the molybdenum usually consisted of the following phases, arranged in succession in depth: $MoSi_2$, Mo_5Si_3, Mo_3Si, and the molybdenum base (Fig. 1). By the method of inert labels and from the characteristic features of the diffusion layer described by V.I. Arkharov [2], it was found that in the Mo-Si system, silicon plays the major role in diffusion, and consequently the formation of a diffusion layer occurs on the inner boundary. Metallographically and by X-ray analysis, it was found that the formation of the molybdenum silicides occurs in the following order:

$$Mo + Si \rightarrow Mo_3Si + Si \rightarrow Mo_5Si_3 + Si \rightarrow MoSi_2$$

at the corresponding phase boundaries, i.e., the formation of $MoSi_2$ occurs at the expense of the lower silicides.

Curves of the isothermal increase in thickness of the Mo_5Si_3 and $MoSi_2$ phases for the temperature 1250°C were plotted (Fig. 2). Analysis of the curves showed that the thickness of the Mo_5Si_3 and $MoSi_2$ phases increased with time according to a parabolic law. Departures from this law were observed only in the initial stage of the process when the rate of formation of the successive silicide phases was greater than the rate of increase in their further diffusion, as determined by the parabolic law. The explanation of this is that the rate of formation of the conditions of the origination and course of this reaction, in particular the temperature, heat of formation, substrate structure, and so forth. The rate of growth of the diffusion layer, however, is determined by the diffusion coefficient, which may be found from the law of increase in the layer thickness. For this it is necessary to know the concentration gradient over the thickness of the layer.

Fig. 1. Photomicrograph of siliconized molybdenum. From top to bottom are arranged the phases $MoSi_2$, Mo_5Si_3, and Mo_3Si.

The silicon concentration gradient in a layer of molybdenum disilicide for the steady-state process was determined by the X-ray method of layered analysis. The concentration gradient for the Mo_5Si_3 and $MoSi_2$ layer was calculated from the increase in weight of the specimens and the thickness of the layers of the corresponding phases for separate time intervals. The calculated diffusion coefficients of silicon in Mo_5Si_3 and $MoSi_2$ for a temperature of 1250°C had the following values:

$$D_{Mo_5Si_3}^{Si} = (0.04 \pm 0.02) \, 10^{-8} \, \text{cm}^2/\text{sec},$$
$$D_{MoSi_2}^{Si} = (0.61 \pm 0.11) \, 10^{-8} \, \text{cm}^2/\text{sec}.$$

To elucidate the complete diffusion pattern in the Mo-Si system, it is necessary to determine the temperature variation of the diffusion coefficients, on the basis of which it is possible to calculate the activation energies characterizing the physical pattern of the process. V. Z. Bugakov and D. Ya. Gluskin [3, 5] have shown that the following relationship is true of reactive diffusion:

$$D(C_1 - C_2) = D_0 e^{-\frac{Q}{RT}},$$

where D_0 is a temperature independent parameter, while Q is the heat of diffusion. On this basis, we plotted a graph of the temperature dependence in the coordinates $\log(D\Delta C)$ and $1/T$ (Fig. 3).

The experimental points lie satisfactorily on a straight line within the limits of our temperature range.

The calculated activation energies for the diffusion of silicon in Mo_5Si_3 and $MoSi_2$ had the following values:

$$Q_{Mo_5Si_3} = (126,000 \pm 12,000) \, \text{cal/mole},$$
$$Q_{MoSi_2} = (57,600 \pm 6,000) \, \text{cal/mole}.$$

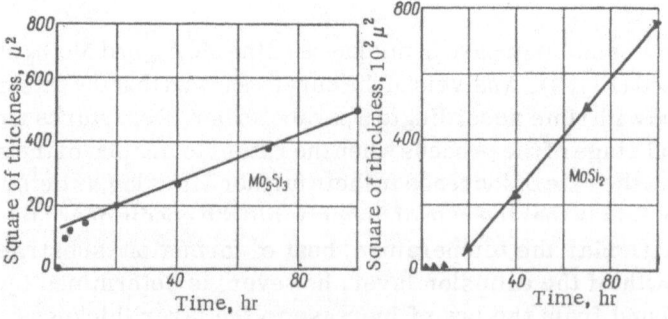

Fig. 2. Dependence of the variation of the square of the thickness of Mo_5Si_3 and $MoSi_2$ layers on time at 1250°C.

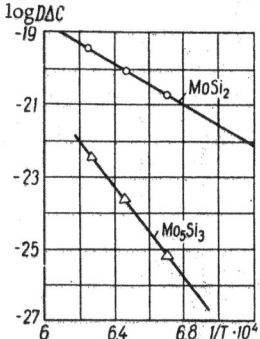

Fig. 3. Dependence of log $(D \Delta C)$ on T for the phases Mo_5Si_3 and $MoSi_2$.

The temperature dependence of the diffusion coefficients of silicon in Mo_5Si_3 and $MoSi_2$ has the form

$$D_{Mo_5Si_3} = 2.2 \cdot 10^8 \exp\left(-\frac{63,000}{T}\right),$$

$$D_{MoSi_2} = 0.8 \exp\left(-\frac{28,800}{T}\right).$$

It should be pointed out that diffusion saturation was conducted in silicon powder. Under conditions such that formation of the metal silicides could have occured as the result of contact of the metal surface with silicon in solid or vapor form, it was necessary to elucidate the mechanism of the interaction of silicon with the metal and, in particular, the function of the vapor phase, the processes of chemisorption and dissociation of the compound formed in the course of the reaction as a whole. In this connection, an investigation was made of the vacuum siliconizing of refractory metals in saturated silicon vapor. Contact of particles of silicon powder with the metal was precluded, the function of the vapor phase was determined during siliconizing, which was carried out in a cell made of solid silicon. Saturation of the silicon vapor in the cell was verified experimentally. Siliconizing of Mo, W, Ta was conducted in the vapor phase, the vapor pressure of the silicon being determined by the experimental temperature. The investigations were carried out at temperatures of 1200 and 1250°C and a residual pressure of 10^{-5} mm Hg.

The curves of the variation of the thickness of the layers of Mo, W, and Ta disilicides with time are shown in Fig. 4.

Analysis of the curves showed a parabolic function of the variation of layer thickness with siliconizing time. Maximum growth rate was found for molybdenum disilicide, and minimum growth rate for tantalum disilicide. The parabolic increase in $MoSi_2$, WSi_2, and $TaSi_2$ was due to the fact that the rate-determining factor in siliconizing was the diffusion of silicon.

In addition to the disilicides, in the diffusion layer the lower silicides Mo_5Si_3, W_5Si_3, and Ta_5Si_3 were found in the form of thin layers between the metal and the top phase. The disilicide layers were compact and of uniform thickness over the entire surface. In the $TaSi_2$ phase, inclusions of the lower silicide Ta_5Si_3 were found. The Ta_5Si_3 crystals were drawn out perpendicularly to the disilicide in the direction of diffusion of the silicon (Fig. 5). Such arrangement of the Ta_5Si_3

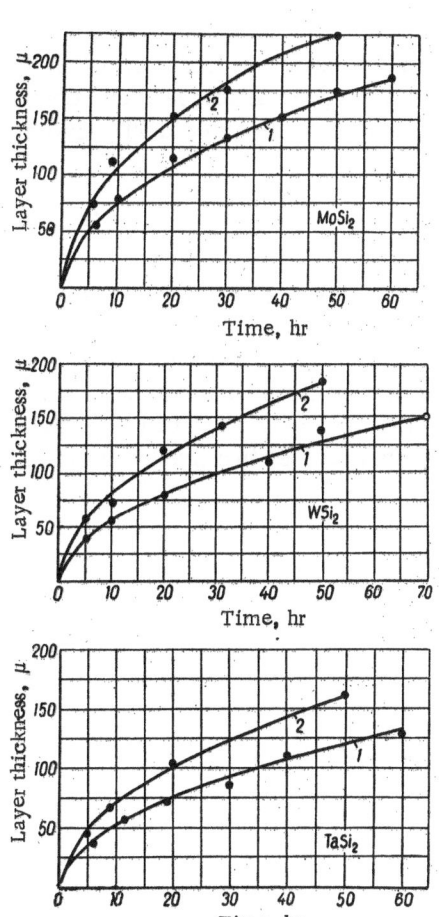

Fig. 4. Variation in thickness of $MoSi_2$, WSi_2, and $TaSi_2$ layers with siliconizing time at 1200 (1) and 1250°C (2).

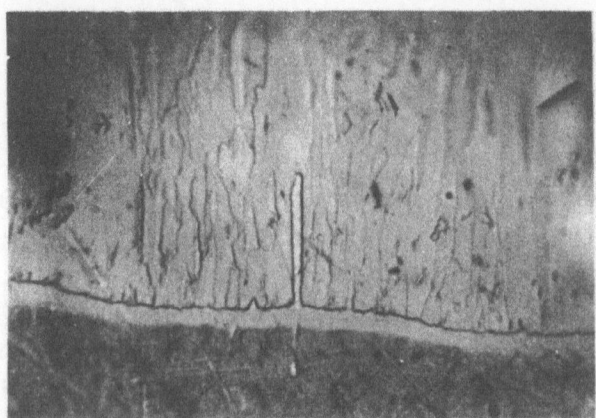

Fig. 5. Photomicrograph of siliconized tantalum
(from top to bottom $TaSi_2$, Ta_5Si_3, Ta; the
Ta_5Si_3 phase penetrates to a considerable distance in $TaSi_2$).

phase gives reason to believe that it is produced within the $TaSi_2$ not as the result of precipitation but is formed on diffusion of the silicon on the $Ta_5Si_3-TaSi_2$ interface.

It follows from the observation made that in the diffusion saturation of tantalum with silicon, lower silicides are first formed, and their transformation to higher silicides then occurs.
The presence of inclusions of the less refractory phase Ta_5Si_3 within $TaSi_2$ is evidently one of
the casues of the pitting of siliconized tantalum in air.

In vacuum siliconizing of the refractory metals Mo, W, Ta, the rate of growth of the diffusion layer of silicides is lower than in siliconizing in the gaseous phase [7, 8].

In siliconizing molybdenum wire by $SiCl_4$ in a current of hydrogen at temperatures of
1200–1800°C, Beidler, Powell et al. [7, 8], found that the rate of growth of the layer increased
with increase in temperature. In this connection it is important to elucidate the influence of
temperature on the specimen in vacuum siliconizing. An investigation was made of the influence
of the temperature gradient between the cell and specimen on the rate of growth of the diffusion
layer on Mo, W, and Ta in a vacuum. The experiments were conducted at two predetermined

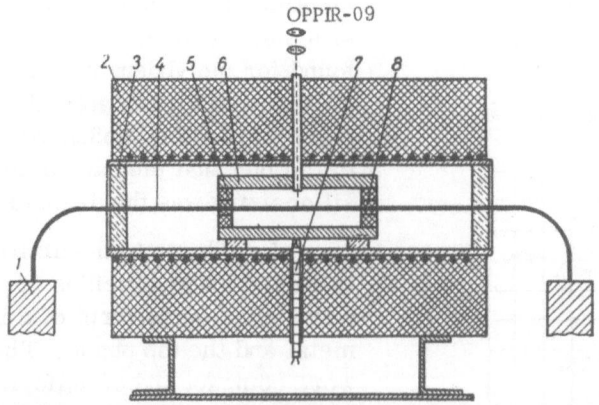

Fig. 6. Diagram of apparatus. 1) Current leads;
2) furnace; 3) furnace cover; 4) specimen; 5) cover
of silicon cell; 6) cell; 7) thermocouple; 8) Al_2O_3
bead.

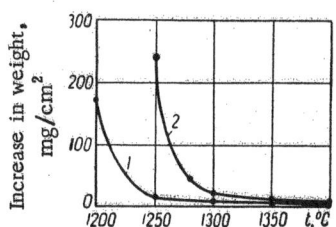

Fig. 7. Variation in weight of a siliconized molybdenum specimen per unit surface with temperature of specimen at cell temperatures of 1200 (1) and 1250°C (2).

cell temperatures of 1200 and 1250°C. The temperature at the specimen in both cases varied up to 1400°C. The apparatus is shown diagrammatically in Fig. 6.

It was found that in the presence of the above-mentioned temperature gradient between the cell and specimen, Me_5Si_3 were formed. Figure 7 shows the variation of the weight of siliconized molybdenum with temperature of the specimen. Siliconizing time was 10 hr.

Figure 8 shows a plot of the variation in thickness of Mo_5Si_3 layer with siliconizing time at a cell temperature of 1200°C. The temperature of the specimen was 1250°C.

It follows from Figs. 7 and 8 that with increase in the temperature of the specimen, the quantity of silicon diffusing into the metal is less, and the growth rate of the silicide layer in the case of a temperature gradient between the specimen and cell is less than under isothermal conditions at corresponding temperature. At the same time, the variation of layer thickness with time also obeys a parabolic law. Consequently, the limiting factor in the siliconizing process in this case, as also under isothermal conditions, is diffusion. For tungsten and tantalum, the results are similar.

The decrease in siliconizing rate in the present case compared with the isothermal siliconizing process is due to the decrease in the silicon concentration gradient with the depth of the diffusion layer. Such a decrease with increase in temperature results from the dissociation of $MoSi_2$ and the re-evaporation of silicon from the surface of the specimen.

It is to be expected that if a temperature gradient opposite to that described above is produced, i.e., if the temperature is less on the surface of the specimen than on the cell, the growth rate of the silicide layers will increase.

Investigations were made of the influence of the temperature gradient between the cell and specimen on the growth rate of an $MoSi_2$ layer for a temperature at the specimen of 1200°C and at the cell of 1250°C. Figure 9 shows the apparatus diagrammatically.

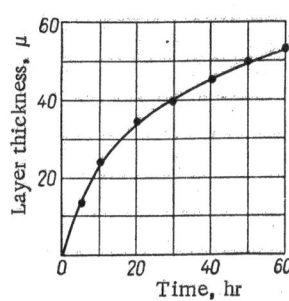

Fig. 8. Variation in thickness of Mo_5Si_3 layer with siliconizing time at a temperature of specimen of 1250°C and cell temperature of 1200°C.

Figure 10 shows the variation of the growth of a layer of $MoSi_2$ with time for a given temperature gradient, from which it follows that the growth rate of the layer in this case is higher than the rate of siliconizing under isothermal conditions, and approaches the rate of siliconizing of molybdenum in $SiCl_4 + H_2$. These experiments show that in the vacuum siliconizing of refractory metals, there is evaporation of silicon from the surface of the saturated metal.

Consequently, the rate of vacuum siliconizing under isothermal conditions is

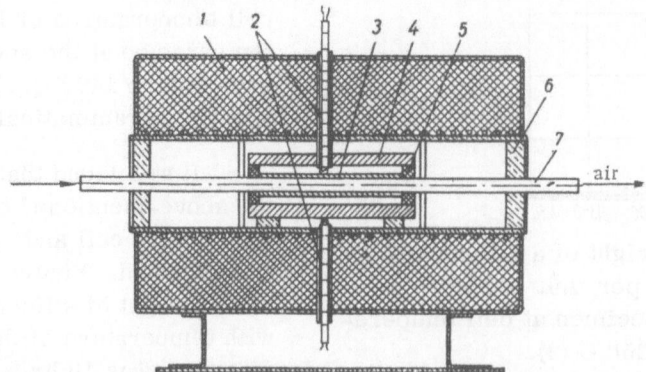

Fig. 9. Diagram of apparatus. 1) Furnace; 2) thermocouple; 3) specimen; 4) cell; 5) Al$_2$O$_3$ ring; 6) furnace cover; 7) stainless steel cooling tube.

lower than the rate of siliconizing in the presence of chlorine. However, as shown by these investigations, it is possible in vacuum siliconizing to obtain rates close to those occurring in siliconizing with SiCl$_4$ +H$_2$.

An activation energy of Q=13,387 cal/mole is given in an article by G. V. Samsonov et al. [6] for the diffusion of silicon in a total silicide layer. The specimens were saturated with silicon in an argon atmosphere from a solid phase bath consisting of 97% Si +3% NH$_4$Cl. The above-mentioned activation energy is approximately one-quarter of that for the diffusion of silicon in MoSi$_2$ in vacuum siliconizing. This may be explained firstly by the fact that in the presence of chlorine, solution of the latter in the MoSi$_2$ lattice occurs, resulting in breakdown of the latter, and consequently in a reduction in the activation energy. The possibility of such a mechanism is pointed out in [1]. Secondly, the evaporation of silicon from the surface of the metal disilicide in vacuum siliconizing leads to a reduction in the rate of growth of the diffusion layer.

CONCLUSIONS

It is found that the dependence of the isothermal growth of MoSi$_2$, WSi$_2$, and TaSi$_2$ layers on the respective metals has a parabolic character.

The activation energy for the diffusion of silicon in Mo$_5$Si$_3$ and MoSi$_2$ has been determined from the temperature curves:

$$Q_{Mo_5Si_3} = (126,000 \pm 12,000) \text{ cal/mole},$$
$$Q_{MoSi_2} = (57,600 \pm 6,000) \text{ cal/mole}.$$

The temperature dependence of the coefficients of diffusion of silicon in Mo$_5$Si$_3$ and MoSi$_2$ has the form:

$$D_{Mo_5Si_3}^{Si} = 2.2 \cdot 10^8 \, \exp\left(-\frac{63,000}{T}\right),$$
$$D_{MoSi_2}^{Si} = 0.8 \, \exp\left(-\frac{28,800}{T}\right).$$

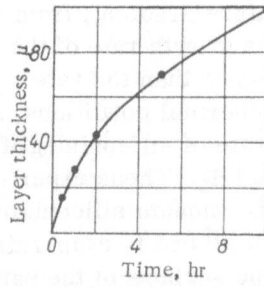

Fig. 10. Variation in thickness of MoSi$_2$ layer with siliconizing time for a temperature of the specimen of 1200°C and cell temperature of 1250°C.

In the case of siliconizing of metals in a cell with a temperature gradient, the rate of siliconizing decreases with increase in the temperature of the specimen, and increases with decrease in temperaure of the specimen in comparison with the cell temperature, while the variation of the growth of the layer with time, for a given temperature gradient, is described by a parabola.

LITERATURE CITED

1. M. I. Aliev and G. B. Abdulaev, Fiz. Tverd. Tela 1:8, 1296 (1959).
2. V. I. Arkharov, Fiz. Metal. Metalloved. 8:2, 193 (1959).
3. V. Z. Bugakov, Diffusion in Metals and Alloys, Gostekhizdat, Moscow (1959).
4. L. F. Verkhorobin et al., Fiz. Metal. Metalloved. 13:77 (1962).
5. D. Ya. Gluskin Zh. Tekhn. Fiz. 20:222 (1950).
6. G. V. Samsonov et al., UkrSSR, No. 1 (1959).
7. E. A. Beidler et al., J. Electrochem. Soc., 98:1 (1951).
8. E. Fitzer, Berg- und Hüttenmaenn. Monatsh. Montan. Hochschule Leoben, 81:97 (1952).
9. S. Mazumi and A. Toshimasa, J. Japan Inst. Metal 21:579 (1957).

BOROALUMINIZING OF IRON AND STEEL

G. V. Zemskov and N. G. Kaidash

Diffusion boriding and aluminizing are widely used in practice for imparting special properties to surface layers on iron and steel [2, 7]. Drawbacks of boride layers are their high brittleness and low heat resistance. Aluminizing increases the temperature resistance and corrosion resistance of iron alloys, but the diffusion layers have low hardness and wear resistance. The saturation of the surface of iron alloys with both these elements is of great theoretical and practical importance. The diffusion saturation with boron and aluminum results in the formation on the iron and steel of surface layers of high hardness and high resistance to oxidation at high temperatures.

Boroaluminizing of steel has been carried out in a gaseous atmosphere [8] formed as the result of the reaction of hydrogen chloride with ferroboral powder at saturation temperatures. It was found that boroaluminizing substantially increased the high-temperature oxidation resistance of steel.

G. V. Samsonov* carried out the simultaneous and consecutive boroaluminizing of steel St 3 and ShKh15 in a mixture of powders of boron carbide, borax, ferroaluminum, and ammonium chloride. He found that in simultaneous boroaluminizing diffusion layers of greater thickness were formed than in separate boriding or aluminizing. Maximum oxidation resistance was possessed by diffusion layers formed by aluminizing previously borided steel.

Despite the promising nature of boroaluminizing, very little has been published on the quantitative relationships and structures of the diffusion layers.

In our investigation, we studied the influence of the composition of the saturating mixture and of the temperature and duration of the process on the structure and thickness of the diffusion layer in the simultaneous and consecutive boroaluminizing of commercial iron and steels 45 and U8A.

Boriding was carried out in a mixture of boron carbide and borax [5], aluminizing in a mixture of ferroaluminum powder and ammonium chloride. Simultaneous saturation was carried out in a mixture of powders of boron carbide, borax, ferroaluminum, and ammonium chloride (see Table 1).

Figure 1 shows the microstructure of commercial iron after simultaneous boroaluminizing. The structure of the diffusion layers depends on the composition of the saturating mixture. Boroaluminizing in mixture 1 results in diffusion layers consisting of acicular borides.

With increase in the ferroaluminum content of the saturating mixture (to more than 15%), a solution of aluminum and boron in iron is formed in the diffusion layer, in addition to borides (Fig. 1a). With a ferroaluminum content of the mixture of more than 50%, the surface layers

---

*Information letter No. 107, Institute of Cermets and Special Alloys, UkrSSR, 1958.

Table 1. Composition of Saturation Mixture for Simultaneous Boroaluminumizing, wt. %

Mixture	84% B₄C+16% borax	97% FeAl+3% NH₄Cl
1	85	15
2	75	25
3	50	50
4	25	75

consist mainly of solid solutions of aluminum in boron and iron (surface zone) and a small quantity of borides adjacent to the base metal (inner zone) (Fig. 1b). Increasing the ferroaluminum content of the mixture increases the thickness of the outer zone of the diffusion layer and decreases that of the inner (boride) layer.

The microhardness of the diffusion layers after boroaluminizing was measured by means of hardness tester PMT-3 at loads of 50 and 100 cN (boride phases) and 10 and 20 cN (solid solution). The microhardness of the outer zone decreased over the depth of the diffusion layer towards the core of the part, being 450–330 daN/mm^2, due to the diminished aluminum and boron contents with the depth of the layer. The boride phases had a high microhardness of 2200–2600 daN/mm^2. An increase in the microhardness of borides on boroaluminizing was evidently due to their being alloyed with aluminum.

The formation of alloyed iron borides is possible in the boriding of steels containing aluminum [1]. An increase in microhardness of the borides is observed also in the borosiliconizing of iron and steel [6].

The specimens having the highest oxidation resistance were found to be those of commercial iron after boroaluminizing in mixture 4, containing 75% ferroaluminum. In a mixture of this composition, saturation of commercial iron and steels 45 and U8A was carried out at temperatures of 800–1100°C with 50-degree intervals for 1, 3, 6, and 9 hr.

Figure 2 shows the variation in thickness of the diffusion layer with saturation temperature (a) and duration (b). The structure of the diffusion layer varies according to the saturation temperature (Fig. 3). In boroaluminizing above 1050°C in the diffusion layer in the solid solution

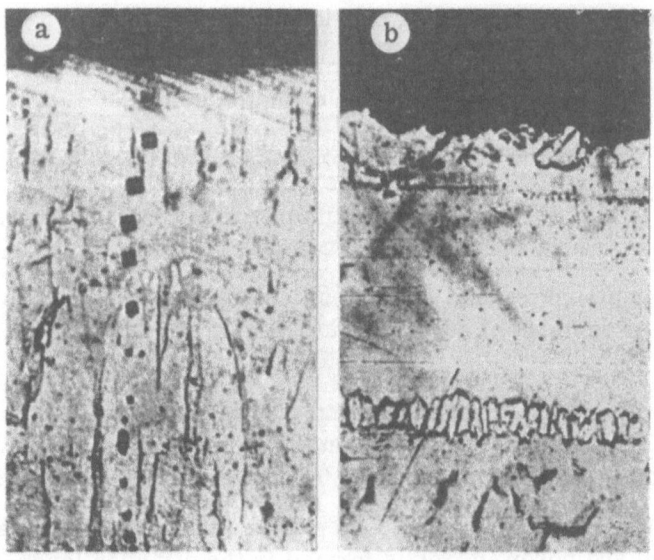

Fig. 1. Microstructure of commercial iron after boroaluminizing in mixture 2 (a), 4 (b) at a temperature of 1050°C for 4 hr. Etchant: 3% solution of HNO₃ in alcohol. (× 340.)

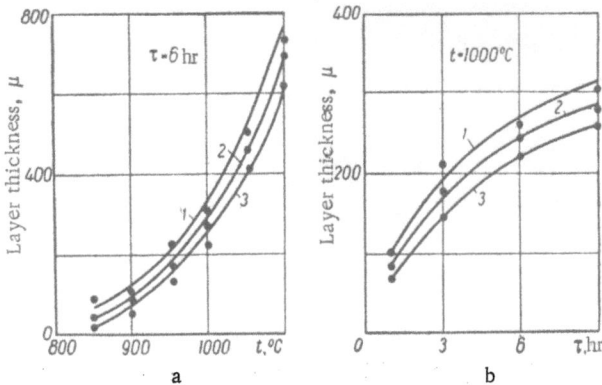

Fig. 2. Variation in thickness of the diffusion layer with temperature and duration of boroaluminizing in mixture 4: (1) Commercial iron; (2) steel 45; (3) steel U8A.

zone, inclusions of a phase of higher hardness occur, evidently aluminum borides AlB_{12} [9] (Fig. 3b). In boroaluminizing of steel, the carbon is driven deeper into the specimen.

Consecutive boroaluminizing was carried out by boriding aluminized specimens, and by aluminizing borided specimens. Figure 4 shows the microstructure of the diffusion layers. In aluminizing a borided specimen of commercial iron, the microhardness in the outer zone of the boride layer is reduced, owing to the replacement of boron by aluminum. In this zone, there are inclusions of higher microhardness, similar to those observed in boroaluminizing at 1050°C. Boride residues are distributed below the outer zone of the diffusion layer. Interlocking of the layer with the base is good, owing to the acicular structure of these residues. Boriding of aluminized specimens above 1050°C evidently results in the formation of aluminum borides AlB_{12}. In this case also, the borides have an acicular structure and are oriented perpendicularly to the saturation surface. The formation of boride phases of high microhardness is also observed at the boundary of the diffusion layer and the saturated base metal. They are formed as the result of the diffusion of boron through the aluminized layer.

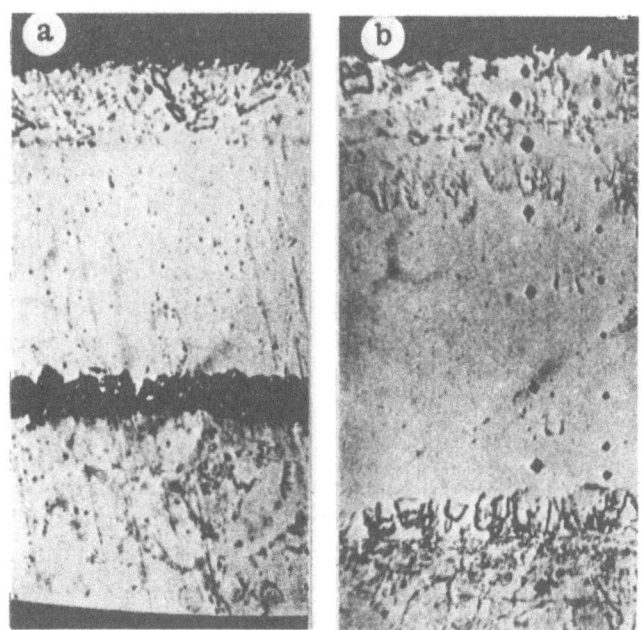

Fig. 3. Microstructure of commercial iron after boroaluminizing in mixture 4: (a) Saturation temperature 1050°C, duration 6 hr; (b) saturation temperature 1100°C, duration 6 hr. Etchant: 3% HNO_3 solution in alcohol. (×115.)

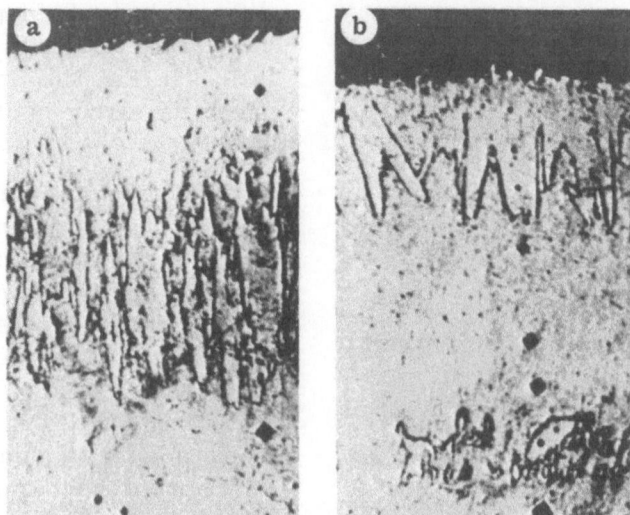

Fig. 4. Microstructure of commercial iron after
boriding-aluminizing (a) and aluminizing-boriding
(b). Boriding at 1100° for 6 hr, aluminizing at 1050°C
for 4 hr. Etchant: 3% MNO_3 solution in alcohol.
(× 115.)

It is important to determine the residual stresses in diffusion layers after combined chemical and heat treatment [4].

Residual stresses in the surface layers are due to the formation in the layer of a structure having a specific volume differing from the specific volume of the saturated metal.

The residual stresses in the surface layer of steel 45 after boroaluminizing were determined in an optical-mechanical apparatus, the layer being removed in a special electrolyte [3]. It was found that residual compressive stresses (-40 daN/mm^2) occurred in the diffusion layer, these compressive stresses changing to tensile stresses at a depth of 15 μ (Fig. 5). Maximum value of the residual tensile stresses was found at a depth of 25 μ (+32 daN/mm^2). The tensile stresses then again abruptly passed into residual compressive stresses. The sharp transition of the tensile stresses into compressive stresses at a depth of 100 μ was caused by the presence of boride phases having a greater specific volume in this zone of the diffusion layer. The compressive stresses attained their maximum value at a depth of 180 μ, and then fell to a minimum value at a depth of 480 μ.

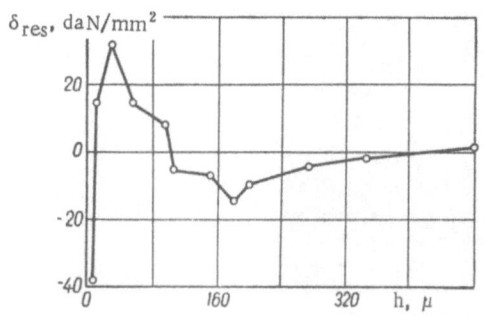

The residual stress method reproduces well the variation in phase composition over the depth of the diffusion layer and successfully supplements other methods employed for their study. It is particularly valuable for studying the layers formed in the diffusion saturation of alloys by a number of elements.

The results of the investigation have shown that in the combined boroaluminizing of iron and steel the structure of the diffusion layer depends on the composition of the saturating mixture. The phase composition of the layer is determined by

Fig. 5. Distribution of residual stresses over the depth of the diffusion layer on steel 45 after boroaluminizing in mixture 4 at a temperature of 1000°C for 6 hr.

the rates of diffusion of boron and aluminum in iron. Due to the fact that boron has a higher rate of diffusion in iron than does aluminium, the growth of the diffusion layer commences with the formation of boride phases. In the boroaluminizing of commercial iron with short holding times, the diffusion layer consists of an external zone of a solid solution, with borides lying underneath. In the case of boroaluminizing at temperatures below 900°C, the formation of boride phases only is found in the layer, which is also evidence that the rate of diffusion of boron in iron is higher than that of aluminum. As the aluminum content in the surface layer increases, retardation of the growth of the boride phase is observed (the width of the boride zone in the layer diminishes). The borides have a more rounded form and extend to a much shallower depth.

Testing of the boroaluminized specimens for high-temperature oxidation resistance was carried out in the temperature range of 700–1050°C. Oxidation resistance was determined from the increase in weight of the specimen. The specimens were weighed on an analytical balance mounted above the electric furnace TG-16, weighing being at equal intervals of time, without removal of the specimen from the furnace. The specimens were placed in small heated porcelain crucibles suspended by a nichrome wire. In each case, a correction was made for the variation in weight of the crucible and suspension. The method of testing boroaluminized specimens for oxidation resistance provided for cycled heating and cooling. At a constant high temperature, the protective oxide film on the specimen is preserved better than in cycled heating and cooling, when the film cracks rapidly, thus assisting the penetration of oxygen to the diffusion coating and the base metal.

Specimens of commercial iron, boroaluminized in mixture 4 and tested at a temperature of 1050°C for 40 hr showed an increase in weight of 0.054 g/cm^2. Aluminized specimens also tested at 1050°C for 40 hr showed an increase in weight of 0.039 g/cm^2. Testing of borided specimens for oxidation resistance above 700°C gave unsatisfactory results. At a temperature of 850°C, the increase in weight in 14 hr was 0.12 g/cm^2.

SUMMARY

A study has been made of the influence of the method, composition of the mixture, temperature and duration of boroaluminizing on the structure and thickness of the diffusion layer on iron and steel.

The layer distribution of the residual stresses in the diffusion layer on steel 45 has been determined. The residual stresses change their sign twice, due to the formation in the boroaluminized layer of phases having specific volumes differing from the specific volume of the saturated metal.

LITERATURE CITED

1.   M. E. Blanter and N. P. Besedin, Metalloved. i Obrabotka Metal., No. 6 (1955).
2.   N. S. Gorbunov, Diffusion Coatings on Iron and Steel, Izd. Akad. Nauk SSSR, Moscow (1958).
3.   L. Gribovskii, Collection "Advanced Scientific, Technical and Industrial Experience," Fil. VINITI, Inst. Nauchn. i Tekhn. Inf., No. P61-126/21 (1962).
4.   G. N. Dubinin and L. Gribovskii, Izv. Vysshikh Uchebn. Zavedenii Chernaya Met., No. 1 (1962).
5.   M. G. Kaidash, Nauk. Zap. Odes'k. Politekhn. Inst., No. 50 (1963).
6.   M. G. Kaidash, Nauk. Zap. Odes'k. Politekhn. Inst., No. 51, (1963).
7.   A. N. Minkevich, Chemical Heat Treatment of Steel, Mashgiz, Moscow (1950).
8.   D. O. Slavin, and N. I. Moskvin, Collection "Materials in Chemical Engineering", No. 6, Mashgiz, Moscow (1950).
9.   W. Hofman and W. Jänische, Z. Metallk. 28:1–5 (1936).

EXPERIENCE IN THE APPLICATION
OF BORIDING IN TRACTOR CONSTRUCTION*

N. S. Zinovich

The heavy wear of parts of the driving tracks of tracklaying tractors and other machines has been the subject of research in the USSR and abroad for many years.

The wear resistance of track pins and links in highly abrasive soils is clearly inadequate. The nature of the wear in an abrasive medium without lubricant depends on the ratio of the hardness of the parts to that of the abrasive. The hardness of mass-produced parts of tractor tracks is much lower than the hardness of the quartz and felspar contained in the soil, and this constitutes a contributory factor to the heavy wear.

One of the advanced methods of increasing the wear resistance of machine parts is diffusion saturation by elements, whereby it is possible to increase the surface hardness of parts to $HV = 1200-2000$. From this standpoint, the diffusion processes of chromizing and boriding appear to be most promising.

For increasing the wear resistance of track links and pins and ascertaining the effect of increasing the hardness of one of the parts of a frictional pair, we investigated the effect of boriding track pins on the life of the parts of a driving track. The investigations were made on commercial track links of type DT-54A tractors, these links being cast in high manganese steel G13L (links of the same type are used on the new tractor models T-75, T-74, DT-75). The commercial pins (steel 50G), rods 22 mm in diameter, 420 mm long, were induction hardened (surface hardness not less than HRC 54, and at a depth of 3.5 mm not less than HRC 45). Pins of the same steel were borided and then high-frequency induction hardened.

The investigation of the wear of parts and specimens cut from them were carried out for the frictional pairs commercial link – commercial pin and commercial link-borided pin.

Electrolytic boriding of the pins was carried out in a bath of molten borax at a temperature of 950°C with simultaneous action of the current. As the result of boriding for 1.5-2.0 hr, a diffusion layer 0.12-0.18 mm deep was produced on the surface of the pins. This layer consisted of the iron borides FeB and Fe_2B,† with a hardness of HV 1400-1900.

---

* The present work was carried out by a group of members of the staff of the Automobile and Tractor Scientific Institute (NATI); the laboratory tests and borating research were conducted by the staff of the Components Research Laboratory under the guidance of Candidate of Technical Sciences R. V. Kugel'; the bench tests by the staff of the Track Systems Laboratory under the guidance of Candidate of Technical Sciences E. G. Popov; the field tests were carried out at the NATI Odessa and Moscow Testing Stations. Electrolytic boriding of pins was carried out in the SKTB MF of Giproneftemash under the guidance of G. I. Yukin.

† The presence of boron carbides in the layer was also possible.

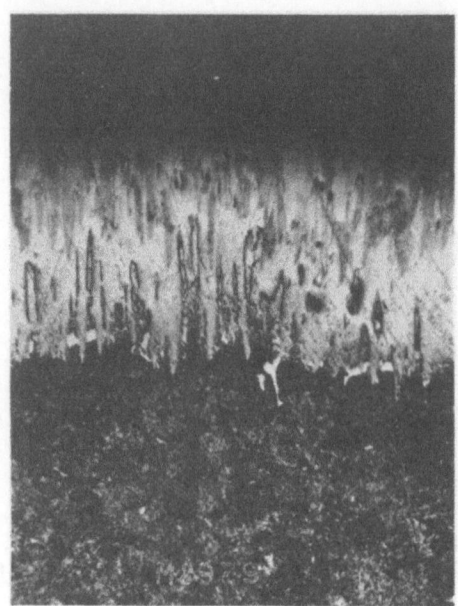

Fig. 1. Microstructure of surface layer of pin, borided, normalized at 830°C for 2 hr, and high-frequency induction hardened (etchant 4% HNO_2). (×250.)

After boriding, the pins were induction hardened on the surface in an apparatus using a valve oscillator.

In view of the fact that the borided pins from the first batch, induction hardened immediately after boriding, were insufficiently strong in service (there were cases of fracture), experiments were made to select the optimum heat-treatment conditions for the pins which would ensure their necessary strength. In particular, normalizing of the pins after boriding was used. Investigations showed that normalizing had no harmful effect on the hardness structure or on the strength of adhesion of the borided layer to the substrate.

The hardness of the borided layer remained within the limits HV 1600-1800, the microstructure of the layer was unaltered, and the structure of the substrate became more disperse as carbon content increased; the microstructure of the core of the pins was considerably improved and had a finer grain, which assisted in improving the quality of the parts.

For further increasing the impact strength of the pins,* induction hardening was followed by oil tempering at 180°C.

As the result of the investigations made, the following treatment schedule was adopted for the subsequent preparation of the experimental batches of pins: boriding at 950°C, 2 hr; normalizing at 820-840°C, 2 hr; induction hardening; oil tempering at 180°C, 2 hr. The pins for laboratory and field tests were prepared according to this schedule.

Examination of pins treated according to the schedule showed the microstructure of the borided layer to consist of columnar crystals of iron borides; the structure of the hardened layer was finely acicular martensite with isolated fine inclusions of troostite (Fig. 1). The transition layer has a martensitic structure, alternating with areas of troostite. The structure of the core was fine-grained and consisted of pearlite and ferrite situated along the grain boundaries. The heat-treatment schedule for the induction-hardened borided parts resulted in an increase of approximately 3-4 times in the impact toughness of the pins.

Testing of the borided pins were carried out under both laboratory and field conditions.

The laboratory tests were carried out by two methods. The first was to test the wear of a pin-lug couple in dry quartz sand on a UIPP−NATI apparatus (Fig. 2). The specimen cut from a pin was inserted in the opening of a specimen cut from a lug, and was rotated in the clamps of the apparatus at a speed of 150 rpm. The specimen from the lug was secured in a device connected by a lever to a weight. Friction of the pin against the lug was carried out under a load of 82.5 N. In tests with an abrasive, dry quartz sand was fed at a rate of 1 liter/hr to the opening of the lug from a hopper.

Borided pins and pins which had not been heat-treated were tested on the UIPP apparatus. Seventy nine pairs of specimens were tested with dry quartz sand and 16 pairs without sand.

_____

*Impact strength was determined by fracture of entire pins on an impact testing machine.

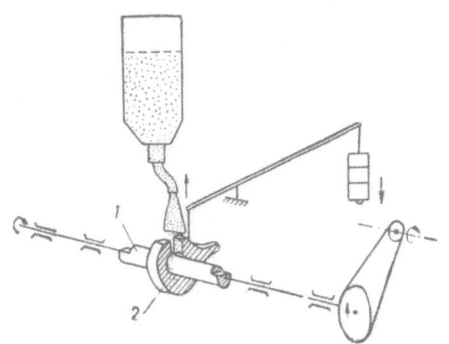

Fig. 2. Diagram of apparatus for testing the wear of specimens: (1) Specimen from pin; (2) specimen from link lug.

For comparison of the wear resistance of the borided pins, parallel tests were made on commercial pins of steel 50G, high-frequency induction hardened to a depth of 3.5-4 mm, and on pins which had not been heat treated. Several series of experiments were made, the total number of tested pairs being 53.

Tests with dry quartz sand were carried out for 4 hr. Wear of the parts was determined by weighing the specimens on an analytical balance after each hour of test. As shown by the test results, 60% of the pins had a wear of 5-7 g; the wear of approximately 90% of the lugs was 2-4 g.

Four series of experiments, comprising 35 commercial pairs with pins having different values, showed that when working in dry quartz sand, variation in hardness of the pins within the limits of HRC 50-64 was without substantial effect on the wear of the parts of a pair (see Table 1). The mean wear of the frictional parts for 4 hr work was 5.50-5.86 g for the pins and 2.67-3.27 g for the lugs. The mean values for the loss in weight (pins 5.8 g, lugs 2.8 g) for the 35 commercial pairs was used as standard for comparison with the wear of other specimens.

The mean wear of the experimental pairs with borided pins for 4 hr was 1 g for the pins (practically 1/6 of the mean wear value of the commercial pins) and 1.1 g for the lugs (1/2.5 that of the mean wear value of the lugs cooperating with the commercial pins).

Figure 3 shows the wear dynamics of the commercial and experimental pins (the dashed lines in the figure were obtained by extrapolation). As can be gathered from the graphs, the mean loss in weight of the borided pins and the lugs cooperating with them for 4 hr was practically the same, being only 1.0-1.1 g, while the wear of the commercial pins and lugs was much higher (the first, six times higher, the second, three times higher).

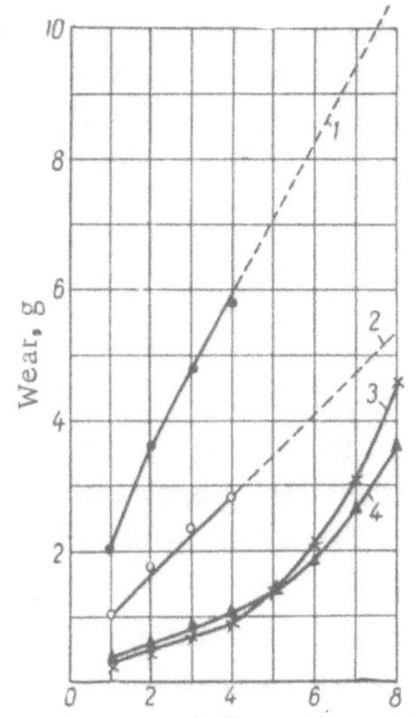

Eight-hour tests on the experimental pairs with borided pins showed that after 5 hr work the intensity of wear on the pins increased rapidly, and the curve of their wear showed a steep rise. The mean wear of the borided pins reached 4.6 g in 8 hr, which was 1.3 times less than the wear of commercial pins operating for only 4 hr. The intensity of wear on the lugs cooperating with the borided pins also increased sharply. After 8 hr work the wear of the lugs increased on the average to 3.6 g, while a similar wear on lugs with commercial pins was observed after only 4 hr.

The total wear of pairs with borided pins after a 4 hr test with sand was relatively low (4.5 times less than that of the commercial pairs), and the loss in weight was evenly distributed between the pins and the lugs.

The results of the laboratory tests showed that under conditions of purely abrasive wear the borided layer not only protected the pin from wear but also

Fig. 3. Generalized wear curves of commercial and experimental pairs tested on the UIPP-NATI apparatus: (1) Commercial pins; (2) lugs with commercial pins; (3) borided pins; (4) lugs with borided pins.

Table 1. Mean Wear of Commercial and Experimental Pairs Tested on UIPP Apparatus for 4 hr as a Function of the Hardness of the Pin

Type of pin	Hardness of pins HV	No. of pairs	Mean wear, g	
			Pins	Lugs
Borided	1600	8	1.00	1.10
Commercial	630—680 (HRC61—64)	16	5.86	2.67
	565—615 (HRC57—60)	10	5.85	2.80
	530—555 (HRC54—56)	4	5.60	2.90
	500—525 (HRC50—53)	5	5.50	2.90

substantially reduced the wear on the lugs. As the borided layer wore away, the rate of subsequent wear of the experimental pairs gradually increased and became the same as the rate of wear of the commercial pairs.

The second series of laboratory investigations consisted of tests of portions of the tracks, comprising 29 links in wet quartz sand from the Lyuborets quarry on stand IG-2V of the Track Systems Laboratory of NATI at track speeds of 8.34 km/hr and a working tension of 110 daN, corresponding approximately to the working conditions of the track on tractor DT-54 in fifth gear with double load (i.e., the same as in second gear).

The duration of the laboratory full-scale tests of the tracks in wet quartz sand was 60 hr. The test were made with commercial and borided pins. The application of borided pins under the given heavy test conditions led to a reduction in the overall wear of the hinges of 5-6 times, a reduction in the wear of the lugs of 2.3-2.5 times, and in the wear of the pins of 8-19 times.

It should be borne in mind, however, that when the duration of the tests was increased beyond 60 hr, the effect of boriding diminished with the wear of the borided layer.

Field tests of commercial and experimental tracks were carried out under normal conditions, with normal operation of the tractors in all forms of agricultural work.

In most of the tracks, both commercial and experimental components were mounted. This made it possible to compare their wear resistance in the limits of each track, thus precluding any error due to possible differences in the dynamic wear of right- and left-hand tracks of the same tractor. The tests were carried out under various conditions of soil and climate, the object being to estimate at least approximately the influence of these factors on the service life of the commercial and experimental tracks.

Owing to the unavoidable scatter of the wear and strength values, a large number of parts were tested, whereby it was possible to obtain reliable criteria, at least for some of the conditions of soil and climate.

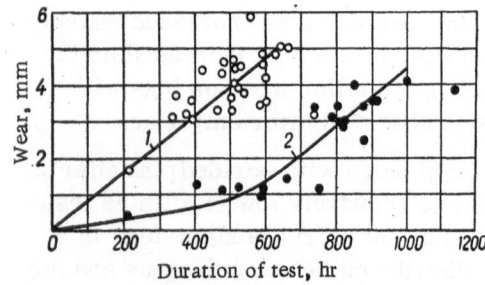

Fig. 4. Wear dynamics of commercial and borided track pins in sandy soils: (1) Commercial pins (25 sets); (2) borided pins (14 sets).

A large number of modifications of the abrasive medium were used in the investigations. For example, wear tests on the UIPP apparatus were made in dry quartz sand, the tests on stand IG-2V were made in wet quartz sand of the same composition; the field tests were made on sandy soils of three districts, on loam soils of two districts, on loess soils and black earth (heavy loam soils). The most complete data were obtained for the wear of pins and lugs in highly abrasive media containing a large amount of quartz sand.

The field tests were carried out in 53 type DT-54A tractors, 12 of which had commercial pins, or mixed sets.

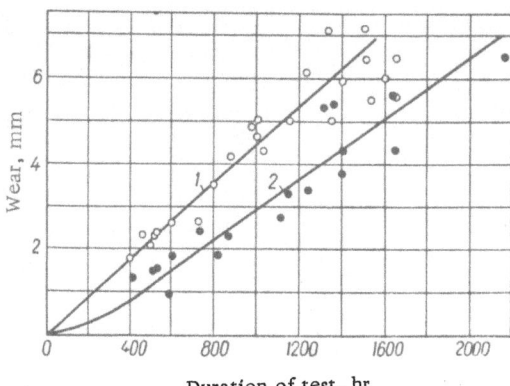

Fig. 5. Wear dynamics of track links in a set with commercial pins (1) and borided pins (2) on sandy soils.

Fig. 6. Wear of borided pin (a) and commercial pin (b) after 667 hr.

The wear dynamics of the commercial and borided pins on sandy soils (Fig. 4) were qualitatively quite similar to the wear dynamics of specimens in laboratory tests on the UIPP apparatus (see Fig. 3). In the initial period while the borided layer was still integral, the rate of wear of the borided pins was much less than that of the commercial pins. When, however, the borided layer had become completely worn (after 500 hr work) the rate of wear of the friction pairs compared became the same (parallel wear curves). The use of borided pins radically altered the wear character of the links.

Figure 5 provides a comparison of the wear dynamics of the lugs of track links in sets with commercial and borided pins in service on sandy soils. Since the wear of link lugs varied with time, depending on the degree of preservation of the borided layer, the degree of measured wear in different testing stages was different; on the sandy soils of the Kherson district, during the entire service life of the tractor two sets of borided pins were used as against three sets of commercial pins.

The use of borided pins substantially retards the increase in pitch of the track, and correspondingly retards the wear of meshing elements (track supporting wheels and driving wheels). The high efficiency of borided pins in the case of tracks used on sandy soils was confirmed by a large number of tests. Figure 6 shows the worn pins from a track after 667 hr service. Whereas the borided pin has scarcely lost its cylindrical shape (its wear is only barely noticeable), the commercial pin shows considerable wear at the points of contact with the link lugs, and the pin itself has the form of a crankshaft, and its unsuitable for further use.

On black earth soils, the average service life of the borided pins was 2400 hr, and that of commercial pins was 1760 hr; the wear of lugs operating in pairs with the borided pins was reduced by an average of 15%.

On loamy soils the service life of the borided pins increased considerably, but at the same time the wear of lugs cooperating with the borided pins was also increased somewhat.

The difference in character the effect of boriding the pins has on the wear of link lugs under different soil conditions is of considerable scientific and practical importance. These differences must evidently be explained by the charging of the worn surfaces with hard particles.

Thus, the application of borided pins in tracklaying tractors DT-54 used on highly abrasive soils increases the service life of the pins by 70-80% (compared with commercial pins), considerably reduces the wear of the link lugs, and retards the wear of the link cogs (due to the retarded increases in pitch of the track.)

This confirms the advisability of producing borided pins for service primarily in tracklaying tractors operating on sandy soils.

STRUCTURE AND PROPERTIES OF STEELS 20KhN3A AND 17N3MA CARBURIZED BY NATURAL GAS CONTAINING ADDED AMMONIA

N. G. Shul'ga, M. M. Fetisova, F. G. Krivenko, and E. M. Tyrman

In drilling for oil and gas, three-roller bits are usually employed. In the working process at a speed of rotation of 500-600 rpm, the bits has an axial load of 300-350 kN. The roller teeth, races, and rolling body of the bearings are subject to extremely rapid fatigue breakdown.

The low strength of the bits, even in the case of correct design and satisfactory machining, is due to the use of steels which are unsatisfactory with regard to chemical composition and mechanical properties, the low hardening level of the thermochemical and heat treatments, and imperfect facing of the rollers with a hard alloy.

In the USSR and elsewhere, much attention is currently being given to questions concerning the search for new types of steel for drill bits and the improvement of the technology of the thermochemical treatment and subsequent heat treatment.

For this investigation, we used the carburizable alloys steels 20KhN3A and 17N3MA. Steel 20KhN3A is widely used for the production of drill bits of different sizes (bit number). Hitherto, hardening of the working elements of the drill bit at all the bit plants in this country has been done by carburizing, using a solid carburizer. The alloy nickel-molybdenum steel 17N3MA, developed and recommended for the production of hits by the All-Union Scientific Research Institute for Drilling Engineering (VNIIBT), has not yet been adequately studied. Specimens of the investigated steels were subjected to carburization with natural gas (from Dashavsk) having a methane content of 98%.

For obtaining the most favorable distribution of carbon over the depth of the diffusion layer and preventing the precipitation of a sooty deposit on the surface of the carburization specimens, a relationship was maintained between the rate of absorption of carbon by the steel surface and the activity of the carburizing gas [1]. The duration of carburization in the active carburizer (1st

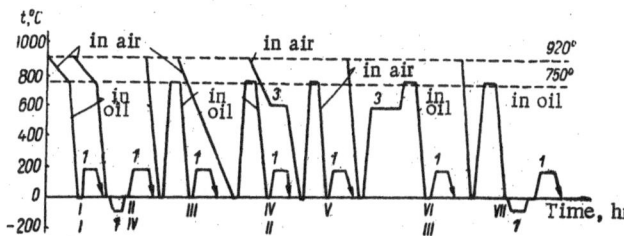

Fig. 1. Heat-treatment schedules of steels 17N3MA and 20KhN3A after gas carburization. The numerals 1 and 3 denote the duration of steps in hours.

49

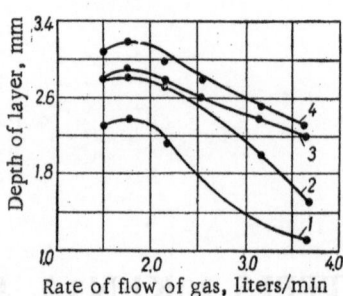

Fig. 2. Dependence of depth of diffusion layer of specimens of steel 17N3MA on the rate of flow of gas and duration of the process at the carburizing temperature 920°C. (1) Without ammonia, holding time 6 hr; (2) with addition of ammonia, holding time 6 hr; (3) without ammonia, holding time 8 hr; (4) with addition of ammonia, holding time 8 hr.

carburizing period) was 25-30% of the total holding time. The supply of natural gas in the 2nd carburizing period varied within the limits of 30-40% of the maximum supply in the 1st period.

For ascertaining the optimum conditions of the thermochemical treatment of the investigated steels, a study was made of the effect on the structure of the diffusion layer of the rate of gas flow in a muffle furnace (carburizer activity), of the addition of ammonia to the carburizing natural gas in amounts of 25-30%, and of the duration of the process.

The effect of the subsequent intermediate and final heat treatment on the structure of the diffusion layer, and on the mechanical properties of the investigated steels was also studied. The specimens were examined for static bending and tensile strength, impact toughness, abrasive wear, wear by friction against metal, hardness, and microhardness.

Heat treatment of the specimens was carried out according to seven different schedules (Fig. 1). Carburization was carried out at 920-930°C.

The rate of flow of the gas and the associated carburizer activity have a substantial influence on the depth of the layer for constant temperature of the process. With increase in the rate of flow of the gas up to a certain limit, the depth of the layer increases, which is due to an increase in the activity of the natural gas used for carburizing. Further increase in the rate of flow of the gas leads to a reduction in the depth of the layer, owing to incomplete dissociation of the methane at high rate of flow of the gas.

The optimum rate of flow of natural gas in the carburization of steels 17N3MA and 20KhN3A in a laboratory furnace with a muffle volume of 12 dm^3 for an active surface of the specimens of 1800 cm^2 was 1.8/1.0 and 2.5/1.0 liters/min, while the depth of carburization for a 6-hr holding time and a temperature of the process of 920-930°C was 2.4 and 2.2 mm (Fig. 2). The addition of ammonia in amount of 25-30% to the natural gas increased the depth of the layer and the carbon content other conditions remaining the same.

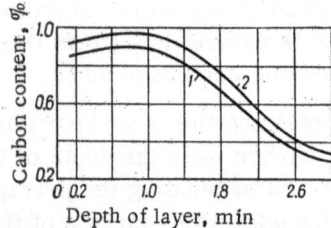

Fig. 3. Distribution of carbon over the depth of the diffusion layer of steel 17N3MA for 8 hr; holding time and temperature of the process of 920°C: (1) Rate of flow of gas 1.8/1.0 liters/min; (2) rate of flow gas 1.8/1.0 liters/min +25% NH_3.

Thus, the maximum carbon content of the diffusion layer of steels 17N3MA and 20KhN3A carburized without the addition of ammonia was 0.9 and 1.1%, respectively, and with the addition of ammonia, it was 0.95 and 1.16%.

The distribution of carbon in the diffusion layer with controlled supply of natural gas is mainly characterized by a smooth transition from the surface (Fig. 3). The presence of ammonia in the carburizer has a favorable effect on the character of the carbon distribution in the layer.

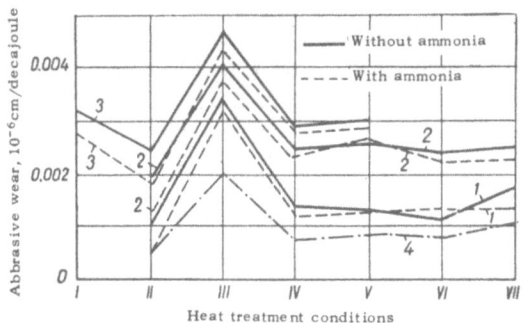

Fig. 4. Influence of ammonia on the abrasive wear of specimens of carburized steel 17N3MA under different conditions of carburization and subsequent heat treatment (rock, limestone). Temperature 920°C: (1) Rate of flow of gas 1.8/1.0 liters/min, 8 hr; (2) rate of flow of gas 3.2/1.5 liters/min, 8 hr; (3) rate of flow of gas 3.7/1.8 liters/min, 8 hr; (4) rate of flow of gas 3.2/1.5 liters/min + 25% NH_3 in the 2nd period (stepped carburization), 8 hr (temperature 920/880°C).

At optimum rates of flow of the carburizing gas and after slow cooling of the specimens in the furnace, the surface zone of the diffusion layer in steels 17N3MA and 20KhN3A has a pearlitic structure with a small amount of excess carbides in the form of isolated inclusions, the amount of which increases in the layer zones situated more deeply from the surface. In the deepest layer zones, the structure is pearlitic and then passes smoothly into the pearlitic-ferritic structure of the core.

With increase in the rate of flow of the carburizing gas, excess carbides in the form of a fine network are observed in the upper zone of the layer.

In the carburization of steels 17N3MA and 20KhN3A with addition of ammonia, the structure of the diffusion layer under optimum gas flow rates is in general similar to the carburized structure, but the transition from the layer to the core is smoother.

After quenching, the surface zone of the diffusion layer of the investigated steels has the finely needle-like structure of martensite with precipitates of fine isolated excess carbides. Below this is a zone with an austenite-martensitic structure which, in the deeper parts of the layer, passes into martensitic structure.

The quantity of residual austenite in the diffusion layer increases with increase in the activity of the carburizer. The conditions of the subsequent heat treatment exercise a marked influence on the quantity of residual austenite. The maximum amount of residual austenite for optimum rate of flow of the gas is found after quenching once, followed by partial cooling and oil cooling. The most effective heat treatment of the carburized steels 17N3MA and 20KhN3A, capable of more complete removal of the residual austenite from the layer, was found to be low-temperature treatment of the quenched specimens at –78°C (schedule II).

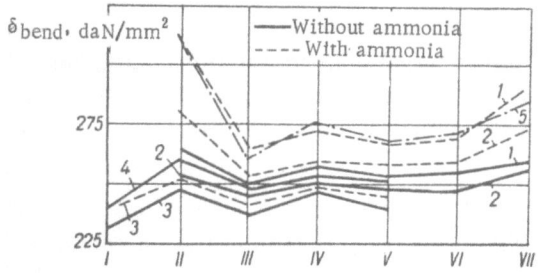

Fig. 5. Influence of ammonia on the bending strength of specimens of steel 17N3MA for different carburizing and subsequent heat treatment conditions (temperature 920°C): (1) Rate of flow of gas 1.8/1.0 liters/min, 8 hr; (2) rate of flow of gas 3.2/1.5 liters/min, 8 hr; (3) rate of flow of gas 3.7/1.8 liters/min, 6 hr; (4) rate of flow of gas 2.1/1.2 liters/min, 6 hr; (5) rate of flow of gas 3.2/1.5 liters/min + 25% NH_3, in period II, 8 hr. (temperature 920/880°C).

The investigation of the abrasive wear of the specimens was conducted on a special apparatus assembled by the Faculty of Metallography and Heat Treatment of the L'vov Polytechnic Institute. The abrasive was limestone from Rudka.

Considerable influence on the wear resistance of carburized specimens of steels 17N3MA and 20KhN3A was exerted by the duration of saturation, addition of ammonia, and also the conditions of the subsequent heat treatment. Maximum wear resistance of the surface of steel 17N3MA was found in specimens which, after diffusion saturation, had been subjected to heat treatment according to schedule II, once quenching and low-temperature treatment at -78°C (Fig. 4). The wear resistance of the specimens carburized with additions of ammonia was appreciably higher than in specimens carburized without additions of ammonia. Similar wear resistance to specimens with heat treatment according to schedule II was shown by specimens treated after carburization, according to schedule IV, with intermediate normalizing. Similar results were also obtained for the wear resistance of steel 20KhN3A, but for all the modifications of thermochemical treatment and heat treatment, the wear resistance of steel 17N3MA remained much higher.

Other mechanical properties of the investigated steels – wear by friction against metal, resistance to static bending, tensile strength, and impact toughness – were also higher in specimens carburized with additions of ammonia at optimum rate of flow of the gas after heat treatments (schedules II, IV). Figure 5 shows the results of static bending tests on specimens of steel 17N3MA.

SUMMARY

Dashavsk natural gas containing about 98% methane is an active carburizer and may be effectively used in the untreated state as gas carburizer in the carburization of drill bit rollers of steels 20KhN3A and 17N3MA.

The optimum rate of flow of natural gas for a constant process temperature (920°C) depends directly on the required concentration of carbon in the surface layer, and on the dimensions of the muffle and of the active surface of the steel (weight of charge). The addition of ammonia in an amount of 25-30% to the natural gas assists in increasing the depth of the diffusion layer and in obtaining a more even carbon distribution in it.

High mechanical properties and wear resistance of the surface were found in specimens of the investigated steels which, after carburization at 920°C with optimum rates of flow of the gas, were subjected to heat treatment according to schedules II and IV. The wear resistance and other mechanical properties of the specimens after carburization with an addition of 25-30% of ammonia to the natural gas were higher than in carburized specimens for all heat treatment conditions.

The most effective influence of ammonia on the mechanical characteristics was found after heat treatment schedule II in which evidently no denitriding occurred. It was proved that the wear resistance of specimens of steel 17N3MA was much higher than in specimens of steel 20KhN3A for all forms of heat treatment.

On the basis of the laboratory investigations of steel 20KhN3A, gas carburization by natural gas with the addition of ammonia for rollers of drill bit No. 12, manufactured by the Drogobyk Machine Construction Plant, has been adopted. The introduction of gas carburization by natural gas with the addition of ammonia and controlled supply of carburizer in the modernized furnace Ts-105 has enabled the quality of the bits to be improved and the labor output to be increased by shortening the carburization time by 50-60%.

LITERATURE CITED

1.   Technical Instruction Data of the All-Union Scientific Research Institute for Drilling Technology, Moscow (1962).
2.   V. G. Chirikov, Podshipnik, No. 7 (1952).

DIFFUSION SATURATION OF STEEL
FROM A GASEOUS MEDIUM
BY HIGH-FREQUENCY INDUCTION HEATING

Yu. V. Grdina and L. T. Gordeeva

The principal feature of the method of high-frequency induction heating is the reduction in the treatment time owing to the acceleration of the processes occurring in the alloys when heated, and the possibility of attaining any temperature. A number of specific features of induction heating are important for diffusion saturation:

Firstly, the component may be heated just to the necessary depth, without affecting the main body of the metal to which optimum structure and properties may be imparted by preliminary treatment; secondly, owing to the high temperature and short holding time, the surface of the metal base may be saturated to the maximum degree, which cannot be done in a furnace since, as the result of diffusion, the alloy layer spreads and its concentration is diminished.

The investigation was carried out using the GZ–46 high-frequency apparatus with a 40KVA, 500-kc tube oscillator. The materials used for the investigations were armco iron, carbon steel (0.64%C) and steel 38KhMA. Saturation was carried out with chromium, aluminum, silicon, tungsten, molybdenum, boron, and also with titanium, followed by boriding. The specimens subjected to investigation had a diameter of 7 mm and a length of 10-15 mm with a machine-finished surface. The specimens were degreased before saturation.

Figure 1 shows diagrammatically the apparatus for diffusion saturation from a gaseous medium.

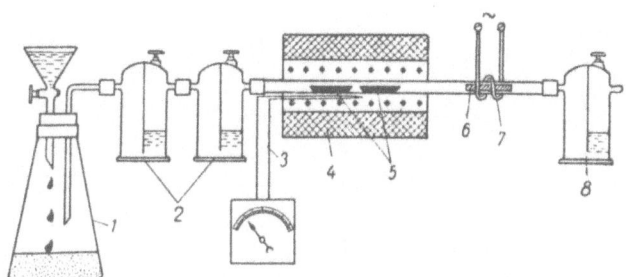

Fig. 1. Diagram of apparatus for diffusion saturation: (1) Flask with $KMnO_4$ and HCl; (2) driers; (3) thermocouple with galvanometers; (4) furnace; (5) boat with metal; (6) specimen; (7) inductor; (8) NaOH seal.

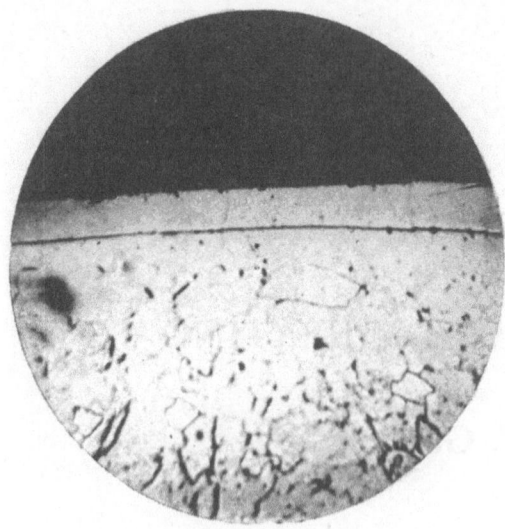

Fig. 2. Siliconized layer on armco iron
(etchant HNO_3), (× 200.)

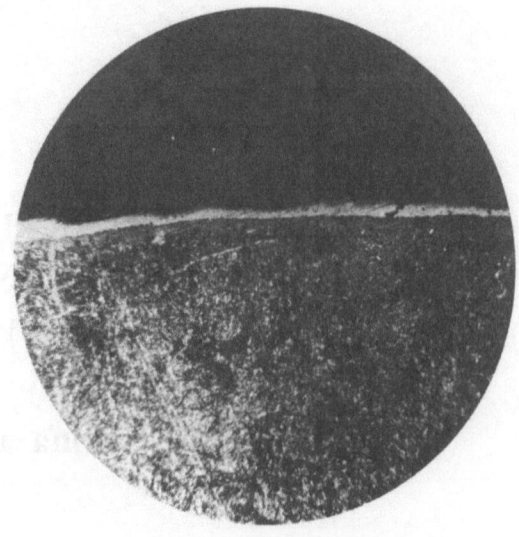

Fig. 3. Aluminized layer of steel 38KhMA
(etchant HNO_3). (× 100.)

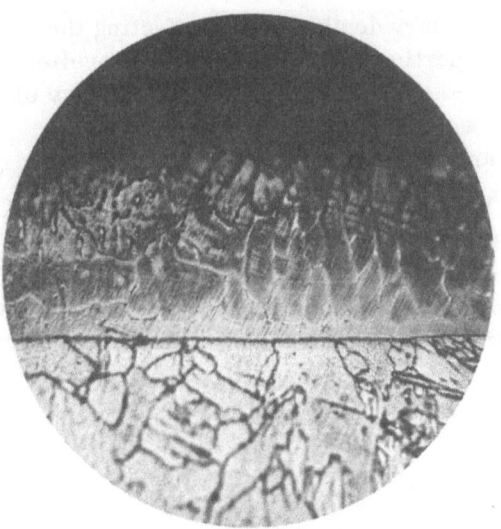

Fig. 4. Aluminized layer on armco iron
(etchant HF + HCl). (× 200.)

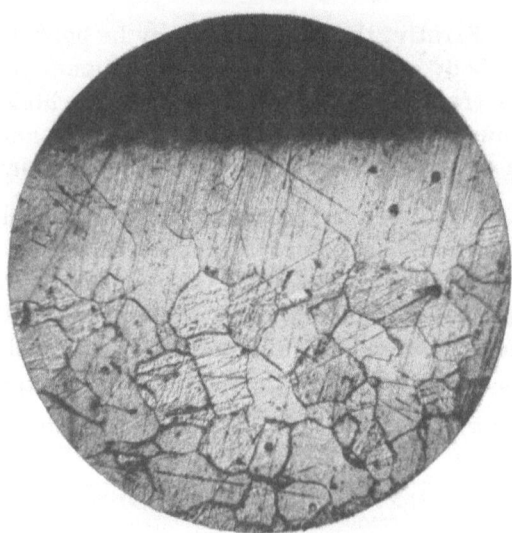

Fig. 5. Molybdenized layer on armco iron
(etchant HNO_3). (× 200.)

The specimen was heated in a quartz tube through which were passed the chlorides of the coating metal. In subsequent experiments, the quartz tube was replaced by a special hermetic chamber with a center driving device for rotating the specimen. Chlorine gas was produced by reaction between concentrated hydrochloric acid and potassium permanganate. For its purification and drying it was passed through Tishchenko flasks containing water and concentrated sulfuric acid. The chlorine, purified and dried in this way, passed along a quartz tube to the furnace, which was heated to the necessary temperature, and contained the metal powder in (preliminarily heated) porcelain boats. The chlorides formed as the results of heating of the metals in the current of chlorine, then passed along the tube and around the specimen placed on "runners" of nichrome or tungsten wire. The powder was thus heated in a flow of

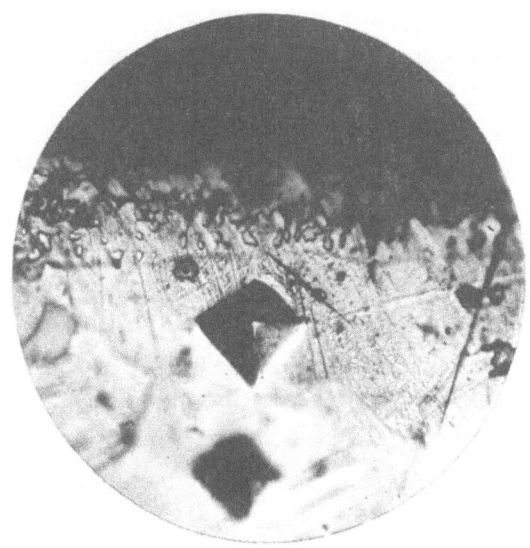

Fig. 6. Boriding layer on armco iron (etchant HNO_3). ($\times$ 600.)

of chlorine and the specimen was also subjected to induction heating in an atmosphere of chlorides in the quartz tube. The inductor was a two-turn coil of copper tubing, 6 mm in diameter; the distance between the inductor and the surface of the specimen was 2-3 mm. At the outlet of the tube was a seal consisting of a Tishchenko flask containing a solution of NaOH for absorbing the chlorine and producing a constant excess pressure of the gas in the tube, the height of the column of liquid being 40-60 mm. To improve the reproducibility of the experimental results and avoid the disintegrating action of the chlorine on the surface of the coated metal and the diffusion coating when formed, the chlorine supply was strictly controlled, and after flushing out and filling the entire system with chlorine, supply of the latter was practically stopped; no more than 20-25 drops of acid was added per minute, and this merely to produce displacement of the chlorides from the zone of their formation to the specimen.

The metals were used in powders in the pure form and in the form of their ferro-alloys of high alloy content for greater completion of the reaction, and were mixed with 5-10% ground chamotte. The reaction for the formation of gaseous chlorides occurred at temperatures of 600-650°C for aluminum, and at 940-960°C for the other metals. Heating of the specimen was carried out by the discontinuous method with a total holding time of 8-30 sec at a temperature of 1000-1100°C (in isolated cases higher). The temperature of formation of the chlorides in the furnace was measured by means of a chromel-alumel thermocouple, while the heating temperature of the specimens was measured by means of an optical pyrometer OPPIR-09. The specimen was cooled in an atmosphere of chlorine and chlorides.

Siliconizing (source of the applied substance was ferrosilicon mark SI75) of aromco iron and carbon steel at a temperature of 1100°C and holding time of 8 sec gave a siliconized layer having a depth of $30-35\mu$ (Fig. 2). The microhardness of the layer as measured in microhardness tester PMT-3 under a load of 1 N was 682, that of the intermediate zone 460, and that of the metal base 356, daN/mm^2.

Aluminizing (source, aluminum powder) was carried out on steel 38KhMA at a temperature of 800-1000°C for 25 sec, depth of layer 20μ (Fig. 3), and on armco iron at a temperature of 1200-1300°C for 8 sec, depth of layer up to 300μ. Figure 4 shows the structure of the diffusion layer after etching with a mixture of the acids HF + HCl. The microhardness of the layer on armco iron was 318 and that of the base 136 daN/mm^2.

Chromizing (source, ferrochrome) was carried out on steel with 0.64%C. Chromium chloride $CrCl_3$ is very heavy and it moved sluggishly over the bottom of the tube. It therefore surrounded the specimen unevenly. The coating layer was formed only from the direction of high chloride concentration (from below). At a temperature of 1000°C and a time of 8 sec, a chromized layer of 0.2 mm was obtained. The microhardness of the layer was 870 and that of the base 345 daN/mm^2.

Tungstenizing (source, ferrotungsten) was carried out on carbon steel and armco iron. At 1100°C and a holding time of 15 sec, the depth of the tungstenized layer for carbon steel was

0.15–0.20 mm. The hardness varied insignificantly; for the layer it was 395 and for the base 366 daN/mm^2. On armco iron for approximately the same depth of layer, the hardness for the layer was 302 and for the base 162 daN/mm^2.

Molybdenizing (source, ferromolybdenum) was carried out on armco iron. The depth of the layer and the variation in hardness were the same as for tungstenizing. The microstructure is shown in Fig. 5.

Boriding of armco iron at a temperature of the chloride-forming furnace of 990–1000°C, temperature of the specimen 1200–1300°C and holding time 68 sec gave a layer having a depth of $14\,\mu$ with a microhardness of the base of 127 and of the layer 1030 daN/mm^2 (Fig. 6).

Titanizing (source, spongy titanium) was carried out on steel having 0.64%C with the formation of a layer of a depth of $12–20\,\mu$ in 20 sec. The microhardness of the metal base was 210 and that of the titanized layer 488–238 $daN/..m^2$.

The experiments have shown that diffusion saturation of iron and steels by the elements Al, Si, Mo, Cr, W, Ti, and B from a gaseous medium (the chlorides of the metals) is possible by means of high-frequency induction heating. Comparison of the depth of the layers showed that the rate of saturation of steel by metals from gaseous media with high-frequency induction heating is several hundred times greater than with heating in a furnace, and the microhardness of the diffusion layers is in the limits indicated in the literature [1].

The properties of the coatings produced on steels were not studied, as this was beyond the scope of our task, which bore the character of a search for the development of a method of producing coatings by induction heating.

LITERATURE CITED

1.   N. S. Gorbunov, Diffusion Coatings on Iron and Steels, Izd. Akad. Nauk SSSR, Moscow (1958).

PHASE TRANSFORMATIONS IN HIGH-TEMPERATURE
GAS CARBONITRIDING (CYANIDING)

V. G. Permyakov and V. G. Tinyaev

The intensification of technological processes is one of the main problems arising in connection with the introduction of total mechanization and automation. It is, therefore, no accident that special attention is being paid to the acceleration of processes for the heat treatment and thermochemical treatment of metals.

An appreciable effect on the intensification of thermochemical treatment processes may be obtained by total action on the saturation process, i.e., by the use of active saturation media advanced methods of heating, and an increase in the temperature at which the process is carried out.

Currently, it may be assumed as established that high-frequency induction heating (or noncontact electrical heating) fails to provide any advantage over furnace heating in regard to the depth and quality of the resulting layer for the same conditions of conducting the process [1, 4]. The advisability of using induction heating and contact electrical heating for thermochemical treatment is determined by the fact that in apparatus with contact electric heating, in contrast to furnaces, the increase in temperature is not limited by the high-temperature oxidation resistance of the working elements, since it is only the treated part which has the high temperature. Contact electric heating still has an indisputable advantage. It enables thermochemical treatment to be used in a mass-production line, which is currently of special importance.

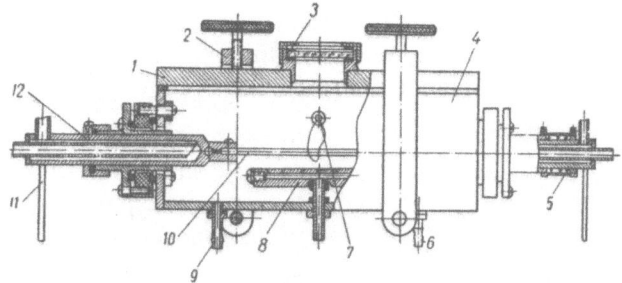

Fig. 1. Gas carbonitriding chamber: (1) Cover;
(2) clamp; (3) inspection window; (4) chamber
body; (5) spring; (6) connection for discharge of
saturating medium; (7) lead-in for thermocouple;
(8) sprayer; (9) connection for the introduction of
saturating medium; (10) specimen; (11) supply
lead; (12) electrode.

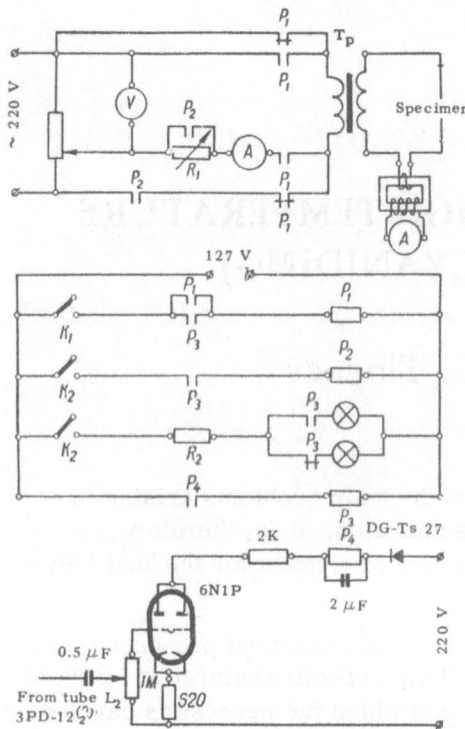

Fig. 2. Electric circuit of gas carbonitriding apparatus.

The rate of thermochemical treatment processes, which, as is well known, are diffusion processes, depends to a considerable extent on the temperature. The application of high temperatures greatly intensifies the saturation process and enables layers of the necessary thickness to be produced in a comparatively short time, estimated in a few minutes, and the subsequent heat treatment ensures high mechanical properties of the diffusion layer.

The purpose of the present investigation was to study the relationships in the formation of a carbonitrided (cyanided) layer at high temperatures with direct electric heating, and the variation in the properties of the layer on subsequent heat treatment.

The starting material used for the production of the carbonitrided layers was armco iron with 0.04% carbon.

For carbonitriding the specimens an apparatus was designed and constructed which contained the following principal components: Gas carbonitriding chamber (Fig. 1); feed and proportioning system for the saturating medium supplied to the chamber; a heating system; and a system for regulating the isothermal heating temperature (Fig. 2). It was possible to carbonitride in the chamber a specimen of a diameter of 1-5 mm and length of 120 mm. Below the specimen was arranged a sprayer for quenching the carbonitrided specimen. The electrodes, in which the specimen was fixed, were watercooled. To obviate bending of the specimen as the result of thermal expansion, one of the electrodes was movable. To prevent leakage of gas from the chamber all the joints of the chamber elements were made with seals of vacuum rubber or Teflon. The gaseous saturating medium supplied to the chamber consisted of a mixture of ammonia and natural gas. The quantity of gases supplied was fed by flowmeters of diaphragm plate type. The specimen was heated by a 4.5 kW power transformer. The system for heating and regulating the temperature of the specimen made it possible to maintain the predetermined isothermal heating temperature to an accuracy of not less than ±3°C (Fig. 2).

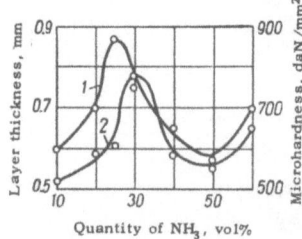

Fig. 3. Dependence of thickness and microhardness of carbonitrided layer on the quantity of ammonia in the gaseous medium: (1) Curve of microhardness variation; (2) curve of layer thickness variation.

As a temperature pickup, a chromel-alumel thermocouple was used which had a high thermo-emf and, as shown by experiment, did not become saturated in an atmosphere of methane and ammonia.

When the apparatus was tested, it was found that in the temperature range of 850-1100°C, the circuit ensured stable regulation of the isothermal heating temperature.

The hardness of the carbonitrided layer was measured on hardness tester PMT-3 under a load of 30 cN. The layered x-ray phase analysis was carried out in a RKD camera, the construction of which had been modified to permit the axis of rotation

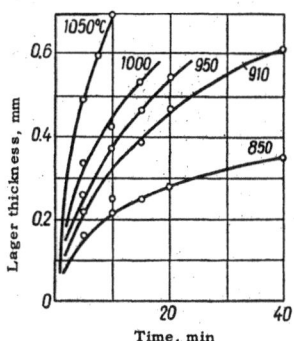

Fig. 4. Dependence of layer thickness on duration of carbonitriding treatment.

of the specimen to be displaced in a plane perpendicular to the primary x-ray beam. By means of this camera, it was possible to obtain x-ray photographs from specimens 1-5 mm in diameter. In photographing, the specimens most frequently used had a diameter of 3 mm. Etching to the required depth was done by means of aqua regia. Tempering of the specimens was carried out in an ordinary tubular resistance furnace in a methane atmosphere. The accuracy of temperature regulation was ±5°C. Tempering time was up to 3 hr.

The causes of the accelerating effect of nitrogen on the carbonitriding process are not yet clear. There is no unified opinion regarding the optimum amount of ammonia in admixture with the carburizing gas for producing a carbonitrided layer of sufficient thickness and possessing high mechanical properties.

In [4, 5] it was shown that a variation in the ammonia content from 2.5 to 10% had the maximum effect on the thickness of the carbonitrided layer. It was pointed out that further increase in the ammonia content of the mixture in direct electric heating sometimes even leads to a reduction in thickness of the layer, and that the ammonia content of the gas medium should be 30-60%.

It appears to us that such contradictory views on the optimum amount of ammonia in the gaseous medium were the consequence of the absence of a strict criterion of the saturation process, such as, for example, the degree of dissociation in the case of "pure" nitriding. In our view, the saturation process will be more completely characterized by the specific rate of flow of the saturating medium, which may be represented as the ratio of the rate of flow of the gas medium to the volume of the reaction chamber and heated surface of the specimen (liter/min · cm^2).

We studied the variation in thickness and hardness of the carbonitrided layer with the ratio of the ammonia and methane supplied to the chamber for a rate of flow of 1.6 liter/min (specific rate of flow 0.2 liter/min · cm^2). For this purpose, we carbonitrided specimens at a temperature of 1000°C and holding time of 15 min. The results presented in Fig. 3 show that maximum thickness and hardness were obtained on the specimens for an ammonia content of the gaseous medium of 25-30%. Subsequently, carbonitriding was carried out at the optimum ratio of ammonia and methane.

For comparison, it is possible to give the value of the specific rate of flow of the saturating atmosphere which may be determined from the data given in [4]. Knowing the rate of flow (0.2 liter/min) and that in an hour the gas was changed 100 times, and also assuming that the heated surface of the specimen was approximately 1 cm^2, the specific rate of flow is found to be 2 liter/min · cm^2, which far exceeds that selected by us. In our view, the introduction of the specific rate of flow makes it possible to compare the results of different investigators, and to state precisely the optimum amount of ammonia in the gas mixture.

Figure 4 shows the dependence of the thickness of the diffusion layer on the duration of carbonitriding in the temperature range 850-1050°C. As was to be expected, the thickness of the layer in isothermal heating varies with time according to a parabolic law; the variation of layer thickness with temperature for constant carbonitriding time is exponential.

Examination of the microstructure of the specimen quenched immediately after carbonitriding as a rule clearly revealed a surface layer of a thickness of 0.02-0.05 mm, consisting of

Fig. 5. Microstructure of surface of carbonitrided layer produced at a temperature of 1050°C and saturation time of 10 min (×300).

carbide and nitride phases. Then followed a layer of martensite and residual austenite, and finally a layer of austenite decomposition products. The martensite was found in layers of a thickness exceeding 0.55-0.6 mm, independently of the carbonitriding temperature. For a layer thickness of 0.55-0.6 mm or more, the concentration of carbon and nitrogen attained a level at which, on subsequent cooling with water, martensitic transformation could occur. Figure 5 shows the microstructure of the surface of the carbonitrided layer, obtained on a specimen at a temperature of 1050°C and holding time of 10 min. The thin layer of carbides and nitrides is followed by a layer with coarse needles of martensite and a large quantity of residual austenite. The quantity of the latter decreases towards the center of the specimen. On passing from the surface, the martensite needles become smaller, and their number increases.

In a study of the microhardness of the carbonitrided layers, it was found that the surface layers possessed the maximum hardness. On passing from the surface, the microhardness fell, this drop becoming more abrupt commencing with a temperature of 950°C. This shows that the concentration gradient of the introduced elements (nitrogen and carbon) in the diffusion layer increased with increase in carbonitriding temperature. In our view, the increase in the concentration gradient of the introduced elements is due to the specific process of saturation in contact electric heating. In the carbonitriding chamber, only the specimen is heated. Dissociation of ammonia and methane is therefore possible only on the surface of the specimen. The surrounding cool atmosphere promotes mixing by convection, the supply to the specimen of fresh portions of ammonia and methane, and the removal of the hydrogen formed. Consequently, in high-temperature carbonitriding in the case of contact electric heating, it is possible to obtain layers with a higher content of the introduced elements than at lower temperatures for the same layer thickness. It is to be assumed that the saturation process is more intense than the diffusion of the introduced atoms to the center of the specimen.

The curves showing the microhardness distribution over the quenched carbonitrided layer obtained at 1000°C and particularly at 1050°C (Fig. 6) in some cases have a rather unusual appearance. On the curves, commencing with a holding time of 5 min, two hardness steps are distinctly visible, one corresponding to a hardness of the order of 850 daN/mm^2, and the second to 750 daN/mm^2.

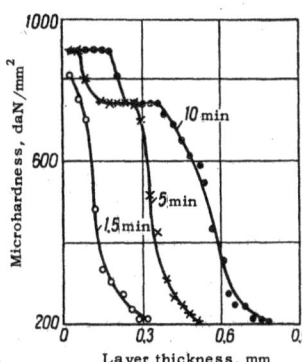

Fig. 6. Curves of distribution of microhardness over the layer, obtained at a temperature of 1050°C and for different carbonitriding times.

With increase in carbonitriding time, the character of the curve does not alter, the layer of maximum hardness merely increases. The dimensions of the second layer scarcely vary, while the layer itself is shifted toward the center of the specimen.

It seems to us that at a temperature of 1050°C, the hardness of the carbonitride layer is determined mainly by the hardness of the carbon-containing martensite, since the nitrogen content of the layer is low. Even the low nitrogen content of the layer produces an increase in hardness (first step). This quantity of nitrogen may be estimated by x-ray analysis. On the x-ray pattern taken from a specimen tempered at 550°C, faint γ'-phase lines are found at a depth of 0.10-0.15 mm. Assuming that the x-ray method is capable of detecting 3-5% of a phase, and that the nitrogen content in the γ'-phase is about 6%, it may be stated with a considerable degree of approximation that the nitrogen content of the layer is 0.2-0.3%, the content being somewhat higher on the surface.

We made a layered x-ray analysis of quenched specimens obtained at a temperature of 1050°C and for different carbonitriding times for the purpose of determining the phase composition across the layer. In every case, the x-ray photographs taken from the surface of the specimen showed lines of residual austenite. The amount of austenite increased with increase in thickness of the layer. On passing gradually from the surface to the center of the specimen, the amount of austenite diminished. At a layer thickness of 0.5 mm or more, in addition to the austenite lines, the x-ray photographs showed martensite lines. The quantity of martensite on passing from the surface at first increased (at the expense of a reduction in the quantity of residual austenite), then reached a maximum, and finally diminished. On passing from the surface, the degree of tetragonality of the martensite diminished.

Tempering of the quenched specimens was carried out in the temperature range of 100-550°C, every 50°C. For studying the effect of tempering temperature on the structure and properties of the carbonitrided layer, the specimens used were obtained at temperatures of 1000 and 1050°C with holding times of 10, 20 and 5, 10 min, respectively. After tempering at each temperature, a microsection was prepared and photographed, and the microhardness was measured across the layer.

The results of the investigation showed that up to a tempering temperature of 150°C, there were no changes in structure. With increase in tempering temperature, etchability of the microsection increased. At a temperature of 250°C, the needles of tempered martensite disappeared. At the same temperature, the microstructure showed no residual austenite. At a temperature above 250°C, the structure was temper troostite.

The microhardness at a depth of 0.15-0.20 mm practically did not change up to the tempering temperature of 150°C, but remained at the level of the quenched layer. With increase in tempering temperature, the microhardness gradually fell from 850 to 420 daN/mm², as was to be expected.

On the basis of the results obtained, it may be concluded that there is no difference in principle in the properties of the carbonitrided layers obtained at temperatures of 1000 and 1050°C if the thickness of the layer is equal to or exceeds 0.55-0.60 mm.

The increase in the efficiency of gas carbonitriding was obtained as the result of the application of contact electric heating, with which high temperatures were easily attained and the dissociation conditions of the saturating atmosphere were varied. The time required to produce a diffusion layer having a high carbon content and containing nitrogen was reduced. By the application of heat treatment, it was possible to modify the structure and properties of the carbonitrided layer over a wide range.

LITERATURE CITED

1. A. D. Assonov, Gas Carburization with Induction Heating, Mashgiz, Moscow (1958).
2. M. M. Zamyatnin and T. A. Balueva, Science of Metals and Heat Treatment, Metallurgizdat, Moscow (1962).
3. Instructions for the Assembly and Operation of EPD-12, Tsent. Byur. Tekhn. Inf. (1961).
4. I. N. Kidin and Yu. G. Andreev, Izv. Vysshikh Uchebn. Zavedenii Chernaya Met. No. 5, (1961).
5. Claude Valat, France-industrie, No. 59 (1961).

APPLICATION OF LIQUID CYANIDING BY POTASSIUM FERROCYANIDE IN MASS PRODUCTION

Ya. N. Funshtein

Liquid cyaniding has become an effective means of increasing the hardness and wear resistance of the surface of parts and tools, and possesses the following advantages: Speed and ease of control of the process; low cost of equipment and production space; possibility of precluding grain regeneration owing to the short duration of the cyaniding process; uniformity of heating of the part; high flexibility of the process, since parts requiring different depths of hardness may be treated in the same bath; and possibility of direct quenching from the salt bath.

For the surface hardening of steel to a shallow depth (0.05-0.5 mm), in modern machine construction cyaniding by cyanides (NaCN, KCN, etc.) is used. Such cyaniding, however, has not been used on a wide industrial scale in machine construction since cyanide vapors are poisonous and harmful to the health of the workmen, and therefore an isolated section must be provided for cyaniding production. As a result, many plants employ gas carbonitriding instead of liquid cyaniding using cyanides. Gas carbonitriding, however, requires special complicated equipment using heat-resistant steel, the duration of the technological process is twice as long or more, and the specific consumption of electric power compared with liquid cyaniding is almost quadrupled.

It is common knowledge that potassium ferrocyanide and ferricyanide are not toxic, and are not dangerous to handle, but unfortunately these salts have not been used in cyaniding at medium and high temperatures.

Data are presented in this article relating to the use of potassium ferrocyanide as principal cyaniding component in liquid cyaniding. In the choice of neutral salts, we took as a basis articles [1, 4], starting from the considerations that these salts had a relatively low melting point (not higher than 520°C) and could be used also for high-temperature tempering (550-680°C) of parts formerly tempered in a lead bath. This made it possible to dispense with expensive baths of lead, which is in scarce supply.

Until recently, it was assumed that the salts entering into the composition of the so-called neutral part of cyaniding baths served only as solvent for the cyanides, and for varying the melting point and increasing the fluidity of the baths. In his work, A. T. Kalinin [1] showed that so-called neutral salts have a paramount effect on the activity of cyanide baths, since they participate directly in reactions determining the rate of the cyaniding process. Baths consisting of sodium cyanide and salts of the alkali metals, sodium chloride and sodium carbonate, have a lower activity than baths consisting of the same sodium cyanide and salts of the alkaline earth metals-barium chloride, barium carbonate and calcium chloride.

Alkaline earth metals improve the washability of salt residues from the parts and increase the fluidity of the bath, while barium carbonate impairs its fluidity, barium carbonate being almost insoluble in water.

64   Ya. N. FUNSHTEIN

Table 1. Data on the Effect of Cyaniding
Temperature and Duration of Holding Time
on Depth of Layer

Cyaniding		Depth of cyanided layer, mm			Hardness after quenching, HRC
Temperature, °C	Duration, hr	Eutectoid	Hypoeutectoid	Total	
820*	1	—	—	0.17	62.0
	2	—	—	0.27	61.5
	3	—	—	0.36	62.0
	4	—	—	0.42	60.0
840*	0.5	0.10	0.07	0.17	57.5
	1	0.14	0.09	0.23	61.0
	2	0.20	0.15	0.35	62.0
	3	0.25	0.20	0.45	60.0
	4	—	—	0.55	
920†	0.5	0.17	0.08	0.25	60.0
	1	0.25	0.20	0.45	62.0
	2	0.33	0.30	0.63	64.0
	3	0.50	0.30	0.80	64.0
	4	—	—	0.91	

*Quenching occurred directly after cyaniding bath.
†Quenching occurred after cooling to 840°C.

The different effects of neutral salts on the activity of cyanide baths is explained by the authors of [4] as being due to the difference in the mechanism of the reactions according to which liberation of active carbon occurs in the presence of a given neutral salt.

On the basis of the foregoing, it was decided to use NaCl and CaCl$_2$ as neutral salts for the cyaniding baths.

After a series of experiments, it was found that the composition by weight of a liquid cyaniding bath should be $^2/_3$ CaCl$_2$ and $^1/_3$ NaCl together with K$_4$Fe(CN)$_6$ (3% of the total weight of the neutral salts). The melting point of this mixture of salts is 520°C.

The experiments were carried out mainly under workshop conditions in type C-100 and type C-60 electrode baths without the use of a steel crucible.

The specimens were prepared from rods 30 mm in diameter or finished pin parts 12 mm in diameter treated (after cold upsetting). The temperature of the bath was measured during the experiments by means of a TKh chromel-alumel thermocouple and was maintained automatically within predetermined limits by a type EPD-12 temperature controller.

For studying the influence of bath temperature and holding time on the depth of the layer, cyaniding was conducted at 820, 840, and 920°C with a holding time of 30 min, 1-4 hr.

The investigation was made on carbon steel marks 15, 20 and 25, 6 to 20 specimens being used of each sort.

After cyaniding, some of the specimens were air cooled, and the others were quenched in water directly from the cyaniding bath, with the exception of specimens cyanided at 920°C. In this latter case, after cyaniding the specimens were transferred to another bath, in which the temperature was maintained with the limits 830-850°C, and they were then quenched in water.

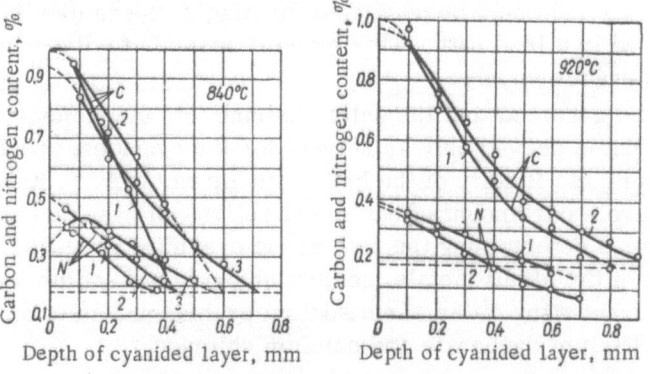

Fig. 1. Distribution of carbon and nitrogen in diffusion layer in the cyaniding of steel mark 20 for different durations of the process.

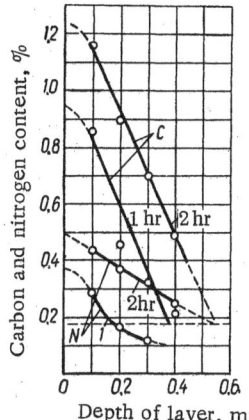

Fig. 2. Distribution of carbon and nitrogen over the depth of layer produced by the cyaniding of control specimens with production parts at 840°C and different durations of the process.

The hardness of the cyanided layer after quenching was measured on "Vickers," "Rockwell," and "Super-Rockwell" type instruments. The data on the effect of temperature and the length of cyaniding on the depth of the diffusion layer are given in Table 1.

As will be seen from the results obtained, the depth of the cyanided layer increased most intensely with increase in temperature and time in the first 60 min, after which its increase becomes slower.

For hardening the surface layer to a depth of 0.1-0.45 mm, it is technologically and economically advisable to carry out cyaniding with potassium (3% concentration) at a temperature of 840°C.

Cyaniding with potassium ferrocyanide at a temperature of 920°C is advisably carried out for a depth of hardened layer of 0.5-1 mm. In this case, it is recommended that the cyanided part be quenched after preliminary cooling in another salt bath to a temperature of 820-840°C.

Such heat treatment of the parts in two baths not only simplifies the technological process, but makes it less laborious, considerably improves the quality of the manufactured parts (absence of decarburized layer, minimum deformation), and reduces the specific electric power consumption.

After quench, the surface zone consists mainly of martensite, and the transition zone of troostite-sorbite.

For determining the carbon and nitrogen contents of the cyanided layer, a layered examination was made of the diffusion layer. For this purpose, chippings were taken to a depth of 0.1 mm from circular specimens of steel mark 20 (C 0.18; Mn 0.5; Si 0.22; Cr 0.05; Ni 0.20%). The specimens were cyanided under various industrial conditions in an electrode bath type C-100. The results of the analysis are shown in Fig. 1.

Table 2. Depth of Layer of Control Specimens Cyanided Together with Production Parts at a Temperature of 840°C in the Bath

Number of control specimen	Duration of cyaniding, hr	Depth of cyanided layer			Hardness after quenching, HRC
		Eutectoid	Hypoeutectoid	Total	
40*	0.5	0.11	0.07	0.18	57
42	0.5	0.10	0.08	0.18	—
77*	0.5	0.13	0.06	0.19	58
30	1.0	0.16	0.13	0.29	—
37*	1.0	—	—	0.22	62
112	1.0	0.14	0.09	0.23	—
20	2.0	0.26	0.20	0.46	—
31*	2.0	—	—	0.50	64
103	2.0	0.22	0.11	0.33	—

*The control specimens were quenched directly from the bath.

For a check on the laboratory and workshop investigations, an addition control check was made of the depth of the diffusion layer in cyaniding production parts of MAZ automobiles.

Cyaniding was carried out in an electrode bath S-100 at a temperature of 840°C. Together with the parts weighing 100-110 kg, check specimens of steel mark 20 (C 0.18%), three for each holding cycle, were charged. One such specimen was quenched directly from the bath, while the others were air-cooled. The depth of layer was checked on all the specimens, and in addition a layered analysis was carried out on one of the specimens for carbon and nitrogen contents.

The results of the control check of the layer depth are shown in Table 2 and Fig. 2.

Figure 3a shows the microstructure of the steel cyanided under workshop conditions.

The chemistry of the cyaniding process using potassium ferrocyanide may be represented as follows. During the melting period of the bath, decompositon of the potassium ferrocyanide takes place in accordance with the reaction [3]

$$K_4Fe(CN)_6 \rightarrow 4KCN + Fe(CN)_2. \tag{1}$$

The presence of calcium chloride evidently ensures an increase in carburizing properties as the result of the liberation of a large quantity of active atomic carbon according to the reactions

$$CaCl_2 + 2KCN \rightarrow Ca(CN)_2 + 2KCl, \tag{2}$$

$$Ca(CN)_2 \rightarrow CaCN_2 + C. \tag{3}$$

The calcium cyanide, $Ca(CN)_2$, and calcium cyanamide, $CaCN_2$, reacting on the surface of the bath with the atmospheric oxygen are decomposed according to the reactions

$$2Ca(CN)_2 + 3O_2 \rightarrow 2CaO + 4CO + 4N, \tag{4}$$

$$CaCN_2 + O_2 \rightarrow CaO + CO + 2N. \tag{5}$$

The carbon monoxide liberated in the decomposition of $Ca(CN)_2$ and $CaCN_2$ according to reactions (4) and (5) is subsequently decomposed according to the reaction

$$2CO \rightleftarrows CO_2 + C, \tag{6}$$

liberating active atomic carbon. Some of the CO finds its way through the body of the slag and burns in the form of yellow tongues of flame. In addition, at the high temperature, $Fe(CN)_2$ decomposes and assists in the simultaneous active saturation of steel by carbon and nitrogen according to the reaction

$$Fe(CN)_2 \rightarrow Fe + 2C + 2N. \tag{7}$$

At a temperature above the Ac_1 point, the atomic carbon dissolves in the γ-iron, while at a temperature below the Ac_1 point, it reacts with iron according to the reaction

$$3Fe + C \rightarrow Fe_3C. \tag{8}$$

The atomic active nitrogen, liberated according to reactions (4), (5) and (7), interacts with α- or γ-iron, saturating the iron and forming nitrides according to the reactions

$$8Fe + N_2 \rightarrow 2Fe_4N, \tag{9}$$

$$4Fe + N_2 \rightarrow 2Fe_2N. \tag{10}$$

During the process of solution of potassium ferrocyanide, a black froth is formed on the surface of the bath, which affords the latter good protection from loss of heat and radiation

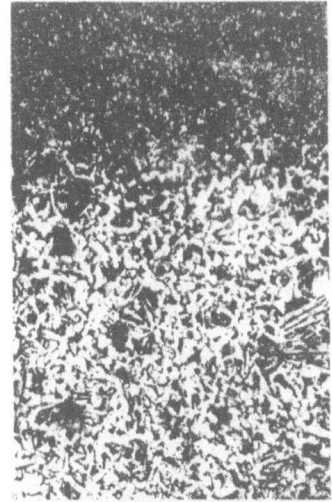

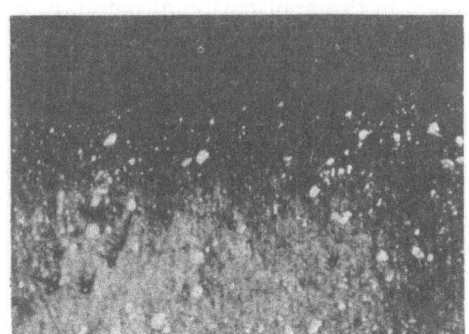

Fig. 3. Microstructure of steel mark 20 after cyaniding in a bath containing $\frac{1}{3}$ NaCl, $\frac{2}{3}$ CaCl$_2$ and 3% K$_4$Fe(CN)$_6$ at 840°C and duration 2 hr (a), and steel mark R18, cyanided in a bath containing $\frac{1}{3}$ NaCl, $\frac{2}{3}$ CaCl$_2$ and 3% K$_4$Fe(CN)$_6$ at 560°C (b).

from the surface of the bath. Investigations carried out with power engineers and technologists of the heat-treatment shops of the Minsk Automobile Plant and the Minsk Motorcycle Plant have shown that the electric power consumption when cyaniding 100-110 kg of parts in electrobath S-100 at a temperature of 840°C for 1 hr was 24 kWhr. (during idle running 19 kWhr), while at a cyaniding temperature of 920°C it was 31-30 kWhr.

The uniformity of the cyaniding results at different levels of the bath was investigated, it being found that the cyaniding capacity of the bath at a depth commencing with 50 mm from the surface of the bath and at the bottom was practically the same. In the same bath, experiments were made to elucidate the degree of exhaustion of the bath as a function of the time. The experiments showed that exhaustion of the bath commenced after three hours' idle running.

For compensating exhaustion of the bath in cyaniding in electric bath S-100, it is necessary to add 1 kg of potassium ferrocyanide every hour.

At the Minsk Automobile Plant, liquid cyaniding of the MAZ automobile is carried out at a temperature of 840 ± 10°C in electrode baths S-100 and S-60 (without metal crucible), containing $\frac{1}{3}$ NaCl, $\frac{2}{3}$ CaCl$_2$ and K$_4$Fe(CN)$_6$ (3% of the total weight of the neutral salts).

Parts of the shock absorber, hand brake, steering wheel shaft supports, speedometer gears, driver's tools, etc., altogether more than 120 items made of steels mark 08, 10, 20, 35, St.3, A12, and 12KhN3A, are cyanided to a depth of 0.05-0.40 mm. The weight of a single part is from 2 g to 15 kg.

Practical experience at the MAZ heat-treatment shop for more than 12 years has confirmed that cyaniding with potassium ferrocyanide under industrial conditions is a completely acceptable, economic and stable process and gives good quality parts.

Liquid cyaniding with potassium ferrocyanide is being successfully employed also at other undertakings of the Belorussian industrial region. An analysis of the cost of the principal materials (reagents and carburizers) and electric power for hardening a layer to a depth of 0.3-0.4 mm showed that cyaniding one ton of parts with potassium ferrocyanide cost 21 roubles 08 kopecks, with sodium cyanide 38 roubles 77 kopecks, and with nitrogen casehardening in furnaces type Ts-60, 33 roubles 26 kopecks; i.e., cyaniding with potassium ferrocyanide is 84% cheaper than cyaniding with sodium cyanide and 60% cheaper than nitrogen casehardening in furnaces Ts-60.

The melting point of the cyaniding bath of the above-mentioned composition is 520°C. We decided to verify the possibility of cyaniding in a bath of the given composition a cutting tool made of high-speed steel.

The experiments were carried out under workshop conditions in electrode bath S-60. The specimens were fully heat-treated cutting and milling tools. Cyaniding was carried out at 560°C. For a holding time of 15 min, the depth of cyanided layer attained 0.025 mm, for 30 min, it was 0.034 mm, and for 1 hr, it was 0.065 mm. Figure 3b shows the microstructure of the cyanided layer of a milling cutter.

Since low-temperature cyaniding by potassium ferrocyanide did not form part of the basic problem of our investigations, but merely bore the character of a check test, more careful investigations are required of the cyaniding kinetics for different sorts of high-speed steels, and of the influence on the red hardness of the tool during cutting operations. The preliminary results, however, give reason to expect that cyaniding of cutting tools of high speed steel with potassium ferrocyanide at 560°C, followed by hot phosphate coating will increase its red hardness.

SUMMARY

On the basis of work carried out and industrial experience in the application of potassium ferrocyanide in cyaniding parts at the Minsk Automobile Plant in the course of more than 12 years, it is possible to draw the following conclusions:

A liquid cyaniding bath consisting of $\frac{1}{3}$ NaCl + $\frac{2}{3}$ CaCl$_2$ + K$_4$Fe(CN)$_6$ (3% of the total weight of the neutral salts) does not represent any danger for the attendant personnel. A salt bath of this composition is universal, since it may be used successfully for both medium- and high-temperature cyaniding and for cyaniding high speed steels, while without or with a slight content of K$_4$Fe(CN)$_6$ it may be used for heating before quenching and for high-temperature tempering.

For hardening the surface layer to a depth of 0.05-0.45 mm, it is technologically and economically advisable to carry out cyaniding in an electrode bath of type S-100 (without the use of a refractory crucible) at 840 ± 10°C with direct quenching, and to attain the desired depth of layer by varying the duration of the process.

The cleanliness of the surface of the cyanided parts and the slight deformation enable cyaniding of parts to be carried out after the final machining operation.

The only drawback in the use of the above-mentioned composition is that calcium chloride is hygroscopic, while when K$_4$Fe(CN)$_6$ is melted a precipitate (iron) is formed which must be regularly removed from the bottom of the bath. Cleansing of the bottom of the bath from deposits and slag should be carried out in each shift (by means of a perforated ladle).

The process of liquid cyaniding with potassium ferrocyanide is flexible, since it is possible to treat in the same bath parts requiring hardened layers of different thickness.

Cyaniding with potassium ferrocyanide in an electrode bath at a temperature of 840°C to a depth of layer of 0.45 mm is a practically and economically acceptable, technologically stable process which can be introduced into any plant without any difficulties.

LITERATURE CITED

1.   A. T. Kalinin, Thermochemical Treatment of Tractor Construction Parts, Mashgiz, Moscow (1954).
2.   S. M. Lyass, Cyaniding and Its Application in Industry, GONTI, Moscow (1939).
3.   A. N. Minkevich, Thermochemical Treatment of Steel, Mashgiz, Moscow (1950).
4.   A. N. Minkevich, Modern Methods of the Heat Treatment of Steel, Mashgiz, Moscow (1954).

SURFACE HARDENING OF TITANIUM BY NITROGEN AND CARBON USING HIGH-FREQUENCY INDUCTION HEATING

Yu. V. Grdina, L. T. Gordeeva, and L. T. Timonina

Research is currently in progress on the surface hardening of titanium and its alloys by means of thermochemical treatment with nitrogen and carbon (separately or together), boron, silicon, molybdenum, chromium, beryllium, aluminum, and oxygen.

Thermochemical treatment of titanium and its alloys is capable of producing a considerable improvement in the operational complex of strength properties: Wear resistance, fatigue strength, resistance to sticking, and resistance to gaseous corrosion.

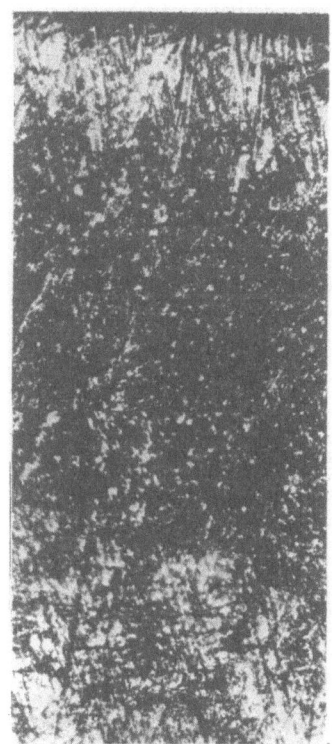

Fig. 1. Microstructure of carbided titanium (etchant HF + HNO$_3$ + glycerin). ($\times$360.)

We have carried out the carbiding and nitriding of titanium alloys by high-frequency induction heating. The advantages of this method of heating for the thermochemical treatment of metals and alloys are given in detail by the authors in [1].

The present article examines the method and results of the surface hardening of titanium and its alloys by carbon and nitrogen. The materials used for the investigation were titanium alloys marks VT-4 and VT-6. The specimens were prepared in the form of cylindrical rods 3 mm in diameter and 200 mm long, and in the form of rollers 40 mm in diameter and 10 mm thick.

The apparatus used for the induction heating was the apparatus G3-46 with a 40 kVA tube oscillator with a frequency of 500 kc. The inductors were made of copper tubing 6 mm in diameter. The surface temperature of the specimens was checked by means of optical pyrometer OPPIR-09.

Carbiding of Titanium. Carbiding in a solid carburizer through CO does not result in the formation of a layer consisting only of TiC; such a process leads to the formation of a mixed layer consisting of TiO$_2$ and TiC. If CO$_2$ is present in the gas, a layer in which the oxide TiO$_2$ predominates is formed. This is confirmed by an investigation [8] which showed that the brittle and coarse layer formed was a solid solution of oxygen in titanium.

We conducted the carbiding of the specimens of titanium alloys by the use of pastes and high-frequency

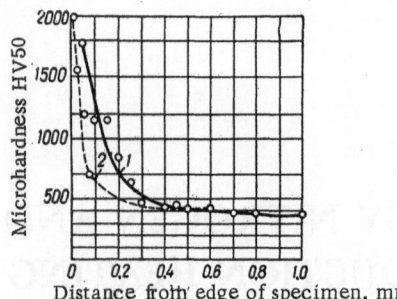

Fig. 2. Variation in microhardness with depth of carbided and nitrided layers: (1) Carbided specimen; (2) nitrided specimen.

induction heating in a helium atmosphere. The use of helium as a control atmosphere precluded oxidation of titanium from the surface and its saturation by harmful gases, causing embrittlement and reduction in the mechanical properties of the metal, such as is the case when nitrogen and hydrogen are used as protective media. In addition, nitrogen combined actively with titanium and a dense nitrided layer is formed on the surface of titanium [2], and therefore a process of saturation with nitrogen, and not with carbon, is more likely to occur. The paste sinters in the form of a crust which peels off and volatilizes. Apart from helium, argon may obviously be used.

The paste of silver graphite and a binder was sprayed on the specimens by means of a compressed air gun. After drying, the specimens were passed into the induction apparatus. Heating was discontinuous in the temperature range 850-1100°C on the surface of the specimens. At this temperature, in 15 min for specimens 3 mm in diameter a carbided layer 0.25 mm deep was obtained (Fig. 1). The maximum microhardness of the surface layer was 1780 HV 50. Figure 2 shows the distribution of hardness with depth.

The roller specimens were tested for wear on a type MI machine at speeds of rotation of the rollers of 220 rpm, load 750 N, with dry friction. A stationary steel roller (U12A), diameter 50 mm, quenched to hardness HRC 45-50, was used as standard. The results of the tests are given in Fig. 3a.

Under the experimental conditions, the rollers carbided for 15 min were found to be practically not subject to wear during a test period of 4 hr (Fig. 3a, curve 3). In a 10-min treatment, the carbide layer produced was very thin and possibly not continuous, and wore out in 20 min. Thickness layers produced in a treatment time of 20 and 30 min (curves 4 and 5) resisted wear well for 2 hr, after which they rapidly started to wear. In the present case, our results are in agreement with the indication [6] that the wear resistance of thin titanium carbide layers is higher than that of thick layers.

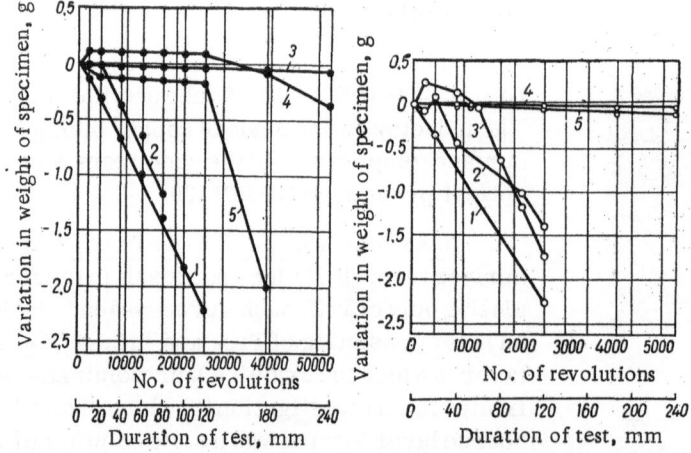

Fig. 3. Testing of carbided (a) and nitrided (b) titanium rollers for wear: (1) Untreated specimens; (2, 3, 4, 5) specimens with duration of carbiding of 10, 15, 20, 30 min respectively, and nitriding 6, 10, 15, and 20 min.

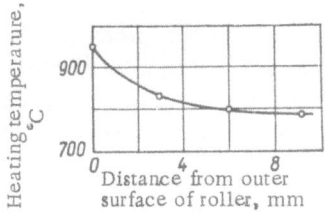

Fig. 4. Heating temperature distribution along the radius of the roller from the surface to center.

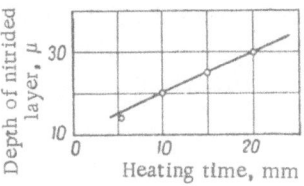

Fig. 5. Variation in depth of nitrided layer with heating time.

<u>Nitriding of Titanium</u>. Investigations on the nitriding of titanium in an atmosphere of nitrogen and ammonia [3, 4, 7] have shown that saturation of the surface of titanium by nitrogen increases the hardness and wear resistance of the surface, and improves its anticorrosion properties at room temperature and elevated temperatures. Long-continued heating in furnaces (up to 24 hr or more) results in a considerable reduction in strength and plasticity of titanium, obviously related to grain growth and saturation of the titanium with hydrogen when nitriding is carried out in ammonia. In our view, it is more rational and effective to carry out the process of nitriding titanium in a nitrogen atmosphere by high-frequency induction heating. High-frequency induction heating enables the temperature of the surface layer alone to be increased, without affecting the bulk of the metal, to which optimum structures and properties have been imparted by preliminary treatment. The use of nitrogen instead of ammonia precludes the possibility of the titanium becoming saturated with hydrogen.

In our experiments, the hermetic chamber for nitriding the specimens was flushed out and filled with purified and dried commercial nitrogen, the rate of flow of which was not measured; its pressure in the chamber was maintained at a level of 40-50 mm water gauge. Heating of the specimens was discontinuous up to a temperature of 850-1100°C on the surface of the specimens. The temperature distribution over the radius of the roller was measured by means of a platinum-rhodium/platinum thermocouple in a drilled-out recess in the body of the roller. Figure 4 shows the temperature distribution in the roller body. A relatively high nitriding temperature was selected for producing titanium nitride of maximum density [5].

The total nitriding time was 6, 10, 15, and 20 min. Maximum thickness of the nitrided layer (up to 30 μ) was obtained in the 20-min treatment (Fig. 5). Figure 6 shows the microstructure of the nitrided layer.

The maximum microhardness of the layer, as determined by means of microhardness tester PMT-3 under a load of 0.5 N, attained 2000 daN/mm^2 (see Fig. 2).

The nitrided specimens were tested for wear under the same conditions as the carbided specimens. The results of these tests are shown in Fig. 3b. As was to be expected, the best results were obtained from specimens with the maximum depth of nitrided layer. Under the conditions described in the foregoing, they showed practically no wear after 4 hr

Fig. 6. Microstructure of nitrided layer of titanium (etchant HF + HNO$_3$). (× 340.)

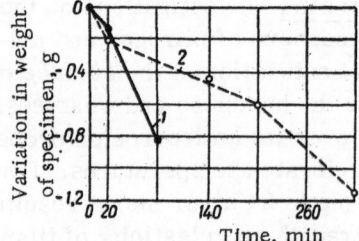

Fig. 7. Oxidation resistance of nitrided titanium at 1000°C: 1) Untreated specimens; 2) nitrided specimens.

of testing: The wear of the untreated specimens was actually much greater than that shown in Figs. 3a and b, since the loss in weight did not include displaced metal left on the roller in the form of deposits on its periphery.

The oxidation resistance of specimens was tested with electrical heating in air on specimens 3 mm in diameter and 200 mm long in a special device made according to drawings of the Central Scientific Research Institute for Ferrous Metallurgy. The temperature was 1000 ± 20°C during the entire test period. The oxidation resistance of the specimens was determined from their loss in weight after mechanical removal of the scale. Testing was carried out to destruction of the specimen. As a result of nitriding, the resistance of titanium to oxidation in air at a temperature of 1000°C (for a thickness of the nitrided layer of 30 μ) was increased by a factor of four (Fig. 7).

SUMMARY

It has been found possible to saturate titanium with carbon and nitrogen by high-frequency induction heating. The depth of the nitrided layer attains 30μ in 20-min treatment; the microhardness of the surface layer is 2000 HV 50. The depth of the carbided layer attains 0.25 mm in 15 min treatment, and the microhardness of the surface layer is 1780 HV 50.

Carbiding and nitriding of titanium considerably increase its wear resistance; nitriding slightly increases the oxidation resistance at a temperature of 1000°C. The optimum depth of the saturated layers ought to be selected for specific wear conditions.

Since in induction heating the core of the specimen is not heated above 800°C (the temperature of the active reaction of titanium with nitrogen), it must be assumed that its mechanical properties are not impaired because of saturation with nitrogen. The use of an induction apparatus with a frequency higher than 500 kc will provide more favorable conditions for the thermochemical heat treatment of the surface of components.

Thus, high-frequency induction heating may be effectively used for the carbiding and nitriding of titanium and its alloys.

LITERATURE CITED

1. Yu. V. Grdina and L. T. Gordeeva, Izv., Vysshikh Uchebn. Zavendenii Chernaya Metal., No. 7 (1959).
2. Yu. V. Grdina et al. Izv. Vysshikh Uchebn. Zavedenii Chernaya Met. No. 6 (1961).
3. A. N. Minkevich, Metalloved. i Obrabotka Metal. No. 7 (1956).
4. E. N. Novikova, Collection "Titanium and Its Alloys," Izd. Akad. Nauk SSSR, Moscow (1960).
5. M. P. Slavinskii, Physicochemical Properties of the Elements, Metallurgizdat, Moscow (1952).
6. A. V. Smirnov and A. D. Nachinkov, Metalloved. i Term. Obrabotka Metal. No. 5 (1960).
7. A. S. Stroev and E. N. Novikova, Collection of Works of the Institute of Metallurgy, Academy of Sciences of the USSR, No. 1, Izd. Akad. Nauk SSSR, Moscow (1958).
8. A. Griest et al., Trans. Am. Soc. Metals 46 (1954).

CHROMIZING OF STEEL BY HIGH-FREQUENCY INDUCTION HEATING IN A VACUUM

G. V. Zemskov and L. K. Gushchin

A drawback in the diffusion metallizing of alloys by various elements is the considerable lengthiness of the process, which can be accelerated only by increasing the temperature. This, however, considerably impairs the properties of the metal base, and also reduces the durability of the furnace equipment at high temperatures.

The thermochemical treatment of the alloy may be accelerated by using induction heating, thereby ensuring a high temperature in the surface layer.

In a number of investigations, diffusion saturation using trowelled-on pastes and powder packs by high-frequency induction heating has been tried [1, 4, 6, 7, 8]. Diffusion coating of steel by metal from a gaseous medium has been found possible [2, 3, 5]. By the use of high-frequency induction heating, the duration of the process is shortened, compared with the use of external sources of heat.

This article describe the investigation of the chromizing of steel by high-frequency induction heating in a vacuum. The chromizing mixture used was a low-carbon ferrochrome mark KhrO and electrolytic chromium. Steels 10 and U8 were used for saturation.

The source of high-frequency current was a tube oscillator GL-15M with a power output of 8.5 kVA and frequency 575-715 kc. The diagram of the diffusion saturation apparatus is shown in Fig. 1.

The temperature of the process was maintained constant by tuning the inductive coupling of the working circuit with the oscillatory circuit, and regulating the mains current of the oscillator tube.

Using specimens 10 mm in diameter and 20 mm in length, the temperature was maintained constant within the limits of 1000-1200°C. The temperature variations did not exceed ±10°C. The surface temperature of the specimen was measured by means of a thermocouple with a wire diameter of 0.2 mm welded to the specimen by the electrical contact method.

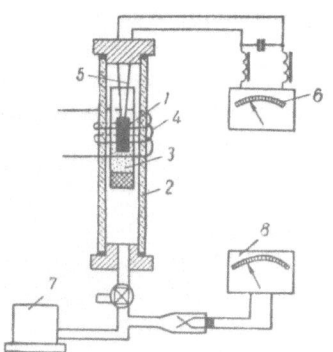

Fig. 1. Diagram of apparatus for chromizing by high-frequency induction heating in a vacuum: 1) Specimen; 2) quartz cylinder; 3) beaker containing saturation mixture; 4) induction coil; 5) thermocouple; 6) galvanometer; 7) vacuum pump; 8) vacuum gauge.

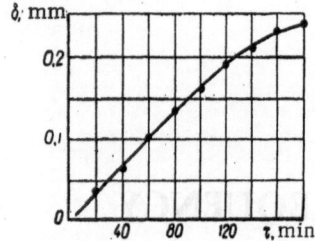

Fig. 2. Dependence of the depth of chromized layer on the duration of saturation of steel 10 at 1000°C in a vacuum in a mixture of 50% FeCr + 50% chamotte.

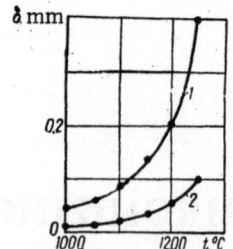

Fig. 3. Dependence of depth of chromized layer on saturation temperature of steel 10 (1) and steel U8 (2) in a mixture of 50% FeCr + 50% chamotte for a holding time of 30 min.

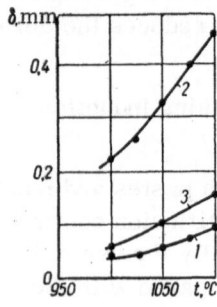

Fig. 4. Depth of the chromized layer obtained with different methods of saturating steel 10: 1) High-frequency heating in a vacuum, mixture of 50% FeCr + 50% chamotte, holding time 30 min; 2) high-frequency heating with the vacuum pump disconnected during the heating, mixture of 50% FeCr + 48% chamotte + 2% NH_4Cl, holding time 30 min; 3) furnace heating in a hermetic container, mixture of 50% FeCr + 48% chamotte + 2% NH_4Cl, holding time 6 hr.

For protecting the galvanometer from currents induced in the circuit, inductances shunted by a capacitance were connected in series with the galvanometer.

During the process of the thermodiffusion saturation of the specimen, saturation of the material of the thermocouple also occurred, and therefore the thermo-emf also varied. The variation in the thermo-emf was investigated for different temperature and holding times. When doing the experiments, corresponding corrections were applied to the temperature readings.

Diffusion saturation of steel with chromium was carried out in a mixture of 50% chamotte in a vacuum of 1×10^{-3} mm Hg, and also in a mixture of 50% ferrochrome, 48% chamotte and 2% NH_4Cl. In the case of saturation with an addition of ammonium chloride, on reaching a vacuum of 1×10^{-3} mm Hg the pump was switched off and heating of the metal took place. Owing to the decomposition of the ammonium chloride, the pressure in the working space rose, amounting to 10–50 mm Hg.

During saturation in a vacuum in the alternating electromagnetic field, intense luminosity of the gases was observed. The thickness of the diffusion layer increase considerably when NH_4Cl was used (Fig. 4). Evidently, during the ionization of the gases and vapors, the course of the surface reactions was accelerated and the supply of ions of the saturaing components was improved by the actions of the "electronic wind."

With the object of verifying the foregoing supposition, an investigation was carried out on the saturation of a steel cylinder closed by a cover. Inside and outside, the cylinder was packed in a mixture of ferrochrome and chamotte. Thermocouples were welded to the inner and outer walls of the cylinder. The thickness of the cylinder walls was 1.5 mm. The temperature of the walls was equalized after heating for a short time. As the result of 30-minute holding time at a temperature of 1000°C, the chromized layer on the outer wall was almost double that on the inner wall. Thus, despite the temperatures being identical, saturation took place less actively on the inner wall of the cylinder. This may be explained by the fact that the walls and cover of the cylinder, acting as screen for the inner surface, weakened the electromagnetic field, and consequently also weakened the ionization in the "electronic wind" in the internal cavity of the cylinder.

SUMMARY

On the basis of the experiments carried out, it has been found that high-frequency induction heating accelerates the process of formation of coatings in a vacuum in powders and with the addition of NH_4Cl. Intensification of the process is achieved as a result of the ionization of the gases and vapor of the metal in the working space and acceleration of the course of the surface reactions. The formation of an "electronic wing" in the alternating magnetic field promotes the supply of fresh portions of reagent to the surface of the metal and the removal of the reaction products.

LITERATURE CITED

1. A. S. Borshcheva and V. S. Gnuchev, Metalloved. i Obrabotka Metal., No. 7 (1957).
2. Yu. V. Grdina and L. T. Gordeeva, Izv. Vysshikh Uchebn. Zavedenii Chernaya Met., No. 7 (1959).
3. Yu. V. Grdina and A. F. Sofroshenko, Izv. Vysshikh Uchebn. Zavedenii Chernaya Met. No. 2, (1963).
4. G. V. Zemskov and I. V. Kosinskii, Nauchn. Zap. Odessk. Politekhn. Inst., Vol. 24 (1960).
5. R. Lorentz and A. Michael, Collection "Problems of Contemporary Metallurgy," No. 4/58 [Russian translation], IL, Moscow (1961).
6. A. N. Minkevich, and G. N. Ulybin, Metalloved. i Term. Obrabotka Metal., No. 4 (1959).
7. I. V. Firger, Vestn. Mashinostr., No. 8 (1951).
8. N. A. Lockington, Metal Forging 24:136 (1957).

THE USE OF REFRACTORY COMPOUNDS IN THE ELECTROCHEMICAL INDUSTRY

V. P. Basov, L. F. Kalinichenko, and A. P. Épik

The combination of many valuable properties, in particular high corrosion resistance and electrical conductivity, of a number of refractory transition metals and their alloys and compounds with carbon, boron, and nitrogen, are opening up wide possibilities for their industrial use. Thus, the problem of the choice of a stable electrotechnical material to be used as current conductor in highly corrosive media is one which is quite important.

From the point of view of availability and industrial use, titanium is quite promising, combining high strength, refractoriness, and low specific gravity with a high corrosion resistance arising due to the presence on its surface of a resistant oxide film formed practically instantaneously on the freshly machined surface. The oxide films retain electron conductivity at the boundary with a conductor of metallic type [3, 4]. However, they also cause a relatively high voltage drop when in electrical contact with certain widely used electrotechnical materials (for example, graphite or mercury), resulting in unnecessary loss of electrical power, overheating of contacts, and other complications, considerably limiting the use of titanium.

If the oxide film on the surface of titanium is replaced by a film of compounds (of the type of carbides, borides and nitrides), which is less resistant to corrosion but of higher electrical conductivity, these limitations are eliminated.

Investigations along these lines have already commenced. Measurements have been made of the voltage drop in the electrical contact of titanium with a carbide, boride, or nitride layer applied by the diffusion method [6], and also the voltage drop of unprotected titanium in contact with various electrotechnical materials. The contact surface of cylindrical specimens (area 5 cm^2) was preliminarily protected and polished. The prepared specimens were brought into contact by clamping under a specific pressure of 10 daN/cm^2. Contact with mercury was effected by immersion of the specimens. The voltage drop was measured at a current density of 20 A/cm^2. The results are given below:

Contact	Drop, mV	Contact	Drop, mV	Contact	Drop, mV	Contact	Drop, mV
Ti–Cu	3	Ti–Cu	2	TiB$_2$–Cu	2	TiN–Cu	2
Ti–Sn	5.5	Ti–Sn	4.5	TiB$_2$–Sn	3.5	TiN–Sn	7
Ti–Ag	7	TiC–Ag	2	TiB$_2$–Ag	1.5	TiN–Ag	1
Ti–Pt	2.5	TiC–Pt	1.5	TiB$_2$–Pt	1	TiN–Pt	0.5
Ti–Fe	5	TiC–Fe	2.7	TiB$_2$–Fe	1	TiN–Fe	5
Ti–Al	7	TiC–Al	4	TiB$_2$–Al	3	TiN–Al	3
Ti–Hg	200	TiC–Hg	90	TiB$_2$–Hg	120	TiN–Hg	60
Ti–graphite	100	TiC–graphite	32	TiB$_2$–graphite	30	TiN–graphite	38
Ti–TiC (carbided layer)	0.5						

77

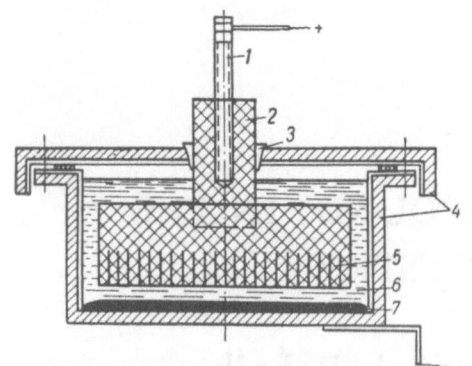

Fig. 1. Diagram of mercury electrolyzer: 1) Metal current supply lead; 2) intermediate graphite rod; 3) seal; 4) rubber coated body and cover of electrolyzer; 5) graphite anode; 6) saturated NaCl solution; 7) mercury cathode.

As can be seen from the above data, the voltage drop in the contacts of titanium specimens, coated with electrically conducting refractory compounds was lower than for the unprotected titanium. The maximum effect was obtained for contact with graphite and mercury. Of the electrically conducting refractory compounds of titanium which were tested, the borided was found to have certain advantages.

It was found that the conditions of applying the electrically conducting compounds affected the value of the contact. Currently, work is in progress on the choice of the optimum conditions for the application of the compounds, whereby minimum voltage drop may be obtained.

The problem under investigation is of great importance in the chlorine industry. It is common that a very aggressive chlorine medium is produced in the electrolysis of chlorides, and the choice of corrosion resistant materials plays a leading part. In modern baths for the electrolysis of chlorides, for example, with a mercury cathode (Fig. 1), the current is supplied to the graphite anode by means of tinned steel or copper rods screwed into an intermediate graphite rod. The metal rods are constantly exposed to penetration of electrolyte, corrosion, and breakdown, accompanied by serious disturbances in the technological process. To prevent penetration of electrolyte to the metal supply leads, the intermediate graphite rods are impregnated with linseed oil or wax. For various reasons, complete insulation of supply leads cannot be achieved in any of the baths. As a result, the iron or copper supply leads serve for three or four working cycles of the anodes (1.5-2 years).

Titanium is a more resistant material in an atmosphere of moist chlorine. Its corrosion figure is 0.127 mm/g [5]. However, on electrical contact with graphite, a considerable voltage drop is observed, which precludes the use of titanium as supply lead [10]. Attempts have been made to protect the contact surface of titanium by electrolytically deposited coatings of copper and tin. Such supply leads have been tested in industrial baths. However, the metal transition films do not adhere well to the titanium base and are subject to corrosion and breakdown on coming into contact with the electrolyte. For this reason, we have made the first experiments in the use of carbided, borided, or nitrided titanium leads.

In addition to the effect of reducing the voltage drop, the corrosion resistance of the coating plays no less an important part, and this factor in turn depends to a considerable extent on the conditions and technology of the application of the coating. If the coating is sufficiently resistant under the given conditions even in the course of the service life of the graphite electrodes (6-7 months), the use of a titanium supply lead becomes quite promising. A titanium rod will last for a very long time (so-called "everlasting" supply lead) with periodical renewal of the conducting coating. This opens up fresh avenues not only for increasing the quality of the products obtained and improving the electrolysis conditions, but for eliminating the impregnation of the graphite rods at all.

Below are given the results of measurements of the voltage drop in a clamped electrical contact of some electrically conducting materials with graphite:

Contact	Drop, mV
Graphite—graphite	35
Graphite—tin.	23
Graphite—copper	27
Graphite—iron	104
Graphite—titanium	100
Graphite—titanium carbide.	32
Graphite—titanium boride	30
Graphite—titanium nitride	38

The measurements were made by the method described above. As will be seen, the electrical contact of graphite with an electrically conducting titanium compound is quite satisfactory, and may replace any of the contacts currently in use.

Corrosion tests were made in specimens of a laboratory model of a mercury electrolyzer in a saturated solution of common salt at industrial current densities (5-10 kA/m^2). Two modifications of the specimens were made: with a clamped contact (Fig. 2a) and a screwthreaded contact (Fig. 2b), an industrial model.

The technology of the preparation of the specimens was as follows: The titanium supply leads were pressed into sockets in the graphite anode with a packing layer of lampblack, and were subjected to carbiding in the assembled form in a graphite tubular resistance furnace. A reference specimen of pure titanium was tested in parallel with the carbided supply leads. Results are shown in Table 1. Tests are currently being made of carbided titanium supply leads under industrial conditions in an operating electrolyzer.

The most important problem in the chlorine industry is the problem of the production of an insoluble anode. The electrographite employed as anode material in universal practice of chlorine electrolysis is an obstacle to the automation and intensification of the technological process. Graphite anodes in use are subject to intense breakdown, causing contamination of the electrolysis apparatus and the products. This results in frequent disturbance of the technological conditions. The application of graphite anodes also involves the need to reduce the interelectrode distance in order to minimize the considerable loss of electric power.

Investigations have recently been made in Russia and elsewhere in a search for new, more effective anode materials [1, 7]. For example, much work has been done in the study of platinum-plated titanium anodes with a titanium supply lead (titanium itself cannot be used as anode material [8], since anodic polarization leads to the formation of a "barrier" layer, caused by a surface oxide layer). Platinum, however, is too expensive a material for extensive use.

Most promising in this direction is the use of titanium and other corrosion-resistant metals coated with an electrically conducting layer of compounds which are stable under conditions of chlorine electrolysis.

We have commenced work on the selection of such anode materials. On a laboratory model, apparatus tests have been made of specimens made from different refractory materials which are stable in corrosive media: Carbides of titanium, zirconium, chromium, molybdenum, and tungsten, and carbided titanium; borides of

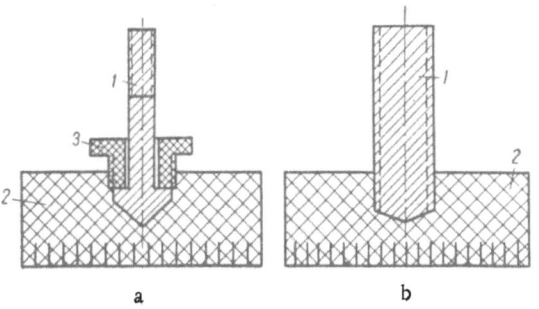

Fig. 2. Specimens of supply leads used in corrosion tests with clamped contact (a) and screwthreaded contact (b): 1) Carbided titanium supply lead; 2) graphite anode; 3) graphite clamping nut.

Table 1. Results of Voltage-Drop Measurements for a Testing Period of 100 hr

Specimen	Voltage drop, mV	
	before test	after test
With unprotected titanium current supply lead	100	105
With clamped carbided titanium current supply lead	62	77
With screwed-in carbided titanium current supply lead	50	48
	14*	17
With screwed-in borided titanium current supply lead	7*	25

*Coating on supply leads was applied separately from the graphite anode.

titanium, zirconium, and chromium, and borided titanium; nitrides of titanium, zirconium, and chromium, and nitrided titanium; and molybdenum silicide. So far, the tests have failed to give the desired results, but this is no reason for discontinuing the investigations.

Evidently, this complex problem will have to be solved not by simple selection but by a careful study of the electrochemical properties of refractory compounds. The promising character of the work which has been started may find confirmation in the fact that two patent specifications [2, 9] have recently been published reporting a resistant electrode which may be used as anode in electrolysis and which consists of a metal (titanium, chromium, niobium) or its alloy, coated with an electrically conducting nitride of the metal.

LITERATURE CITED

1.   British Patents Nos. 905141, 13014660 (1959).
2.   British Patent No. 886197 (1962).
3.   A. F. Ioffe, Semiconductor Physics, Izd. Akad. Nauk SSSR, Moscow (1957).
4.   O. Kubaschewski and B. E. Hopkins, Oxidation of Metals and Alloys, Academic Press, New York (1953).
5.   G. V. Samsonov, High-Temperature Compounds, Metallurgizdat, Moscow, (1963).
6.   G. V. Samsonov, and A. P. Épik, Coatings of High-Temperature Materials, Plenum Press, New York (1966).
7.   L. M. Yakimenko et al., Khim. Prom., No. 10 (1962).
8.   L. M. Yakimenko and I. V. Kokhanov, Khim. Prom., No. 1 (1962).
9.   H. Beet, Canadian Patent No. 643672 (1962).
10. C. Carter, British Patent No. 823598 (1959).

SOME PROPERTIES OF CARBIDE AND BORIDE DIFFUSION COATINGS ON REFRACTORY METALS

A. P. Épik, G. A. Bovkun, I. V. Golubchik, and L. P. Sinitsina

The investigation of the process of the diffusion saturation of refractory metals by various elements has been dealt with in a large number of publications, a bibliography of which is given in [13]. However, only in a few investigations [2, 3, 8, 19] has a study been made of the physico-chemical properties of the diffusion coatings formed, investigations in this direction being of a sporadic character. The further accumulation of experimental material, its assessment, and systematic arrangement are advisable and necessary steps in extending the fields of application of coatings of refractory compounds in engineering.

The present article describes an investigation of the resistance to scaling, wear resistance, and chemical resistance of carbide and boride diffusion coatings on titanium, zirconium, molybdenum, and tungsten, and borided diffusion coatings on niobium. The technology of the production of the coatings is described in [13, 14, 21]. The boride coatings on titanium, zirconium, niobium, molybdenum, and tungsten consisted of the phases TiB_2, ZrB_2, NbB_2, $Mo_2B + Mo_2B_5$, $W_2B + W_2B_5$, and the carbide coatings consisted of TiC, ZrC, Mo_2C, and $W_2C + WC$.

Testing for Resistance to Scaling. Tests for resistance to scaling were carried out with the object of obtaining semiquantitative results showing the effect of carbide and boride coatings on the oxidation process of refractory metals, and also of ascertaining the temperatures at which the oxidation processes of the borided or carbided metals are only incipient, or, on the contrary, proceed very intensely. For comparison with the carbided and borided specimens, specimens of the pure metals were also oxidized.

The published data accumulated on the resistance of scaling of metal-like refractory compounds [15-18, 20, 28, 29, 30] show that carbides of the refractory metals, as a rule, are less resistant to high-temperature oxidation than the borides of the corresponding metals. This is explained [15, 16] by various mechanisms of the oxidation of these refractory compounds. In the oxidation of carbides, in addition to the oxides of the metal, gaseous oxides of carbon are formed, loosening the oxide film and thereby impairing its protective properties. In the oxidation of borides, in addition to oxides of the metal, boric anhydride is formed, which is volatile at relatively low temperatures; at higher temperatures, however, pyroborates of the refractory metals may be formed, and these compounds increase the protective properties of the oxide film.

According to the data of [7, 18, 20, 27-29, 32] titanium carbide commences to oxidize slightly even at temperatures of 500-600°C, and fairly actively at 800°C and above. Zirconium carbide shows approximately the same behavior. It is interesting to note than at 1000°C, titanium carbide, however, oxidizes much more slowly than pure titanium [32].

81

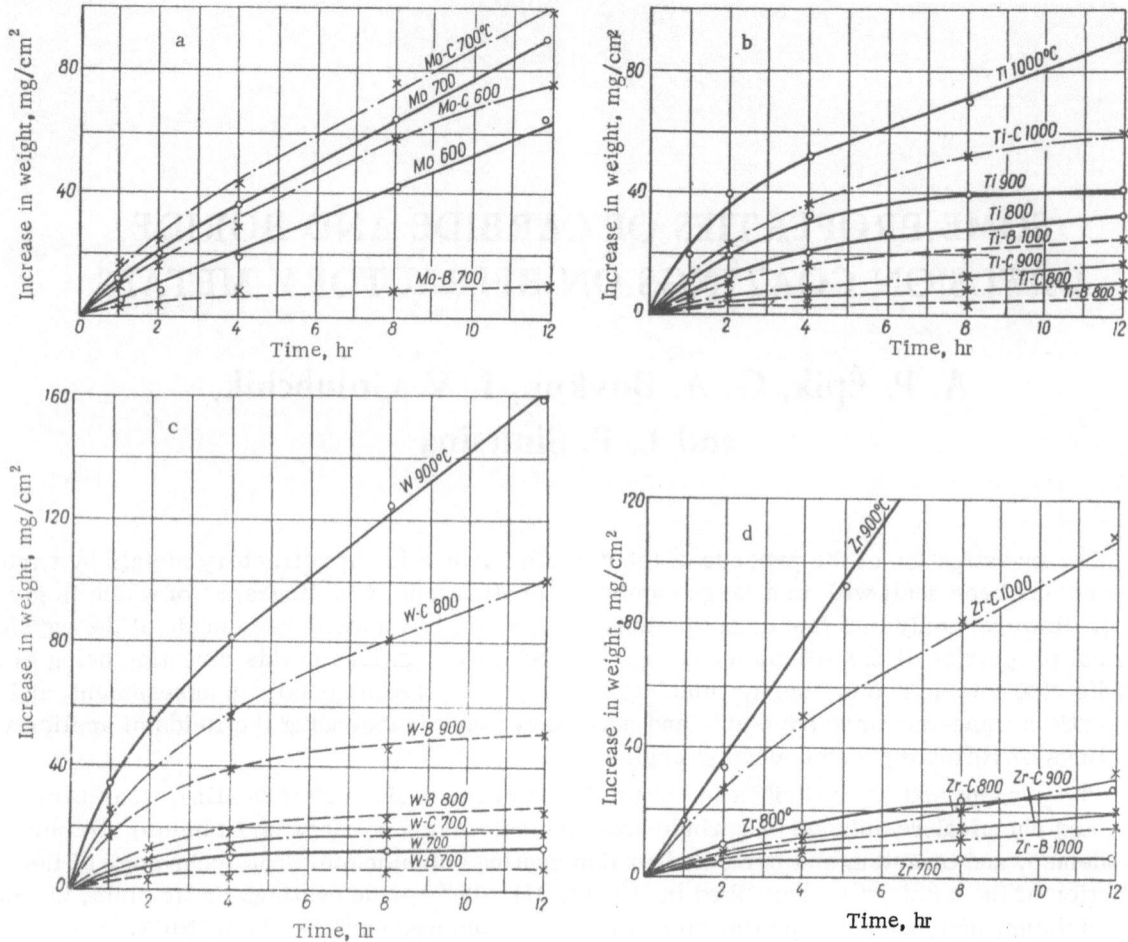

Fig. 1. Rate of oxidation of initial carbided and borided (a) molybdenum, (b) titanium, (c) tungsten, and (d) zirconium at different temperatures in air.

At temperatures of 500–600°C, the carbides of tungsten and molybdenum oxidize much more rapidly than the carbides of titanium and zirconium, and quite intensely even at 700–800°C. Throughout the entire temperature range of 700–1000°C, the tungsten carbides W_2C and WC are oxidized more intensely than pure tungsten [32]. According to [20], the carbides of the refractory metals may be arranged in the following order of increasing resistance to scaling:

$$Mo_2C < WC < W_2C < VC < TiC < TaC\,(NbC) < ZrC.$$

Of the borides, the most resistant to scaling are titanium and zirconium diborides, noticeable oxidation of which commences only at temperatures of 900–1000°C [7, 27, 28]. Niobium and tantalum diborides oxidize fairly rapidly even at 900–950°C [30]. According to [2], a diffusion coating of NbB_2 on niobium is actively oxidized at 1000°C. The oxidation of molybdenum and tungsten borides proceeds intensely at lower temperatures, while according to [30], of the molybdenum borides, Mo_2B is oxidized most actively of all, MoB less actively, and Mo_2B_5 still more slowly. The borides may be arranged in the following order of decreasing resistance to scaling [16]:

$$TiB_2 > ZrB_2 > NbB_2 > TaB_2 > W_2B_5.$$

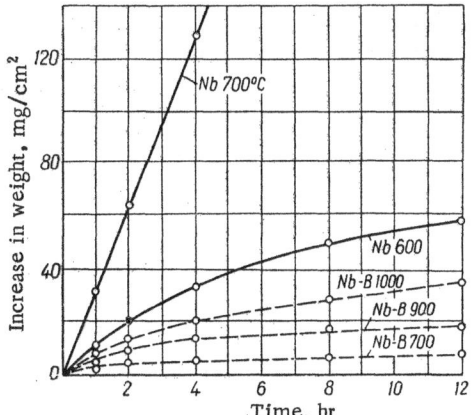

Fig. 2. Rate of oxidation of pure and borided niobium at different temperatures in air.

Taking into account the slight thickness of the diffusion coatings and published data on resistance to scaling of carbides and borides, testing was carried out at temperatures of 600-1000°C, and with a holding time of 1-12 hr.

The results of the tests of the initial carbided and borided specimens, carried out in parallel under identical conditions, are given in Figs. 1 and 2. In every case, boride coatings possessed a higher resistance to scaling than carbides on the corresponding metals. Carbided specimens of tungsten and molybdenum oxidized more intensely than the pure metals [32], carbided molybdenum (which, owing to the high volatility of MoO_3 was tested at a temperature not exceeding 700°) oxidizing particularly intensely.

The specimens of borided tungsten and molybdenum possessed better scale-resisting properties than the initial metals or the carbided metals. Thus, borided tungsten oxidized much more slowly than the pure metal at 900°C, and the carbided metal at 800°C.

The oxidation of carbided and borided specimens of titanium and zirconium became noticeable only at temperatures of 700°C or above, the pure metals in every case possessing a much lower resistance to scaling than the carbided and especially the borided metals. These results are in good agreement with the data of [32]. The difference becomes particularly noticeable at a temperature of 800°C and above. The increase in weight of specimens of carbided titanium at

Table 1. Increase in Weight of Specimens Tested
for 12 hr at Temperatures of 600-1000°C

Specimen	Increase in weight, mg/cm^2, at temperature, °C				
	600	700	800	900	1000
Specimens of the pure metals					
W	6	15	27	—	—
Mo	60	90	—	—	—
Ti	0.3	0.9	34	42	88
Zr	0.2	5	30		—
Nb	5.8		—	—	—
Carbided specimens					
W	2.4	22	100	—	—
Mo	70	97	—	—	—
Ti	0,1	0,48	10	19,6	60
Zr	0,2	0,6	22	35	105
Borided specimens					
W	0,3	4	20	55	—
Mo	0,5	5	—	—	—
Ti	0	1,2	5	9,6	25
Zr	0	0,6	5	9,6	20
Nb	1	7,9	11	32	40

Table 2. Results of Wear Tests on Specimens

Material of specimens	Weight wear ΔQ, g	Volume wear Δv, cm^3	Diameter of specimen d, cm	Area of worn surface S, cm^2	Linear wear Δl, mm	Reduced wear $\Delta l'$, mm	Factor by which wear resistance increased
			Tungsten				
Uncoated	0.0264	0.001370	0.199	0.0310	0.432	0.426	1.0
Carbided	0.0016	0.000085	0.200	0.0310	0.027	0.027	15.8
Borided	0.0010	0.000050	0.203	0.0320	0.015	0.0155	27.4
			Molybdenum				
Uncoated	0.0347	0.003400	0.0152	0.0181	2.000	1.150	1.0
Carbided	0.0024	0.000240	0.154	0.0180	0.133	0.078	14.5
Borided	0.0008	0.000080	0.153	0.1840	0.043	0.0257	46.0
			Titanium				
Uncoated	0.0150	0.003400	0.186	0.0270	1.250	1.090	1.0
Carbided	0.0017	0.000400	0.190	0.0830	0.140	0.125	8.7
Borided	0.0008	0.000190	0.188	0.2780	0.068	0.060	18.2
			Niobium				
Uncoated	0.0302	0.003600	0.151	0.0178	2.000	1.140	1.0
Borided	0.0003	0.000030	0.151	0.00177	0.020	0.0114	100.0

800°C with a holding time of 12 hr was 3.5 times less than the increase in weight of specimens of the pure metal. Under these conditions, the resistance to scaling of borided specimens of titanium and zirconium increased by a factor 6-7 compared with the original metal.

The resistance to scaling of niobium when borided was appreciably increased. On the pure metal, rapid formation of an oxide film occurred even at 600°C, and this film cracked and peeled off the base. At a temperature of 700°C and holding time of 8 hr, a specimen of pure niobium completely disintegrated. Borided niobium began to oxidize at 600°C, while fairly intense oxidation was observed only at temperatures of 900-1000°C [2]. Table 1 gives the values for the increase in weight of specimens of the tested materials after a holding time of 12 hr at different temperatures. As can be seen from the table, the highest resistance to scaling was shown by specimens of zirconium and titanium having on their surface the phases ZrB_2 and TiB_2, and the lowest resistance to scaling was shown by carbided specimens of tungsten and molybdenum with the phases $W_2C + WC$ and Mo_2C on the surface. The tested boride coatings may be arranged on the following order of decreasing resistance to scaling:

$$ZrB_2 > TiB_2 > NbB_2 > W_2B_5 > Mo_2B_5,$$

while the carbides may be arranged in the order

$$TiC > ZrC > W_2C(WC) > Mo_2C,$$

which is in good agreement with published data on the resistance to scaling of the corresponding carbide and boride phases.

Wear-Resistance Tests. The borided, carbided, and pure metals were tested for wear resistance under conditions of abrasive wear, since the literature has no information not only on abrasive wear resistance of coatings based on refractory compounds, but also on the actual compounds of the refractory metals. In the meantime, the high hardness of carbide and

Table 3. Results of Corrosion Tests on Borided Tungsten, Molybdenum, Titanium, Zirconium and Niobium on Boiling in Different Media (Weight before test, after test, and weight lost, g)

Medium	Tungsten	Molybdenum	Niobium	Titanium	Zirconium
HNO$_3$ Concentrated	1.7209 1.7182 0.0027***	5.8832 5.8073 0.0759***	3.1763 3.1696 0.0067**	1.8316 1.8065 0.0251**	0.4490 0.4299 0.0191**
HNO$_3$ (1:1)	1.8940 1.8923 0.0017***	—	3.2083 3.2036 0.0047**	1.7933 1.7661 0.0272**	0.3668 0.3508 0.0160**
H$_3$PO$_4$ Concentrated	1.6931 1.6915 0.0016***	4.6201 4.6174 0.0027***	1.1033 1.1009 0.0024**	1.8691 1.4525 0.4166**	0.3559 0.2971 0.0588**
H$_3$PO$_4$ (1:1)	1.6792 1.6777 0.0015***	4.8501 4.8486 0.0015***	1.1339 1.1330 0.0009**	1.8398 1.8368 0.0030**	0.3089 0.3009 0.0080**
HCl Concentrated	1.4690 1.4678 0.0012***	5.1707 5.1695 0.0012***	1.0819 1.0804 0.0015**	1.8911 1.6780 0.2131**	0.4328 0.4254 0.0074**
HCl (1:1)	1.5046 1.5036 0.0010***	5.4660 5.4650 0.0010***	—	1.7044 1.5247 0.1797**	0.3740 0.3680 0.0060**
H$_2$SO$_4$ Concentrated	1.2762 1.2520 0.0242***	4.4044 3.3118 1.0926***	3.5795 2.4345 1.1450*	1.8624 1.8020 0.0604**	0.4369 0.1084 0.3285
H$_2$SO$_4$ (1:1)	1.5357 1.5337 0.0020***	4.6825 4.6803 0.0022***	3.6615 3.6480 0.0135**	1.7772 1.2397 .5375**	0.4157 0.3979 0. 278
KOH 30%	1.7145 1.7062 0.0083***	4.3999 4.3686 0.0 13***	—	1.7152 1.7120 0.0032**	0.4175 0.4144 0.0031
KOH 10%	1.3394 1.3347 0.0047**	—	—	—	0.317 0.316 0.001

*Testing time 1 hr.
**Testing time 2 hr.
***Testing time 3 hr.

boride phases and consequently of coatings based on them may be used for the protection of various parts used under conditions of abrasive wear. The resulting accumulation of factual data in this field may be of practical interest.

The experimental data acquired may also be used for establishing the relationships between wear resistance of refractory compounds and their hardness, brittleness, structure, and other important characteristics.

M. M. Khrushchov and M. A. Babichev have developed a method of testing abrasive wear and have examined the abrasive wear of many materials, including pure metals, alloys, and minerals [22-26]. The process of abrasive wear of commercially pure metals under different conditions has also been studied by other authors [4, 6, 11]. In this work [4, 6, 11, 20-26] interesting relationships have been obtained between hardness, elastic modulus, and structural composition of the materials on the one hand and their wear resistance on the other.

Table 4. Results of Corrosion Tests of Carbided Tungsten, Molybdenum, Titanium and Zirconium on Boiling in Different Media (Weight before test after test, and weight lost, g)

Medium	Tungsten	Molybdenum	Titanium	Zirconium
HNO₃ Concentrated	2.1491 2.1502 +0.0011*	—	1.6779 1.6752 0.0027***	0.3204 0.3162 0.0042**
HNO₃ (1:1)	2.0907 2.0931 +0.0024**	—	1.8400 1.8392 0.0008**	0.3623 0.2602 0.1021**
H₃PO₄ Concentrated	1.7603 1.7600 0.0003**	5.8454 5.8455 +0.0001***	1.8577 1.0685 0.7892*	0.3444 0.2888 0.0556**
H₃PO₄ (1:1)	2.0233 2.0230 0.0003***	5.7441 5.7438 0.0003***	1.8516 1.8028 0.0488**	0.3181 0.3177 0.0004**
H₂SO₄ Concentrated	2.2748 2.2636 0.0112*	—	1.7668 1.6778 0.0890**	—
H₂SO₄ (1:1)	2.1254 2.1246 0.0008*	—	1.8398 1.0380 0.8018**	—
HCl Concentrated	2.2360 2.2356 0.0004***	5.9717 5.9717 0.0000***	1.7429 1.4400 0.3029**	0.3253 0.3111 0.0142**
HCl (1:1)	2.4617 2.4615 0.0002	5.6744 5.6739 0.0005***	2.1067 1.8630 0.2437**	0.3318 0.3226 0.0092**
KOH 30% 30%	1.7924 1.7922 0.0002***	4.7106 4.7104 0.0002***	1.8255 1.8245 0.0010**	0.3285 0.3283 0.0002***
KOH 10% 10%	1.8067 1.8067 0.0000	—	1.8119 1.8116 0.0003	—

*Testing time 1 hr.
**Testing time 2 hr.
***Testing time 3 hr.

Testing for abrasive wear was carried out on the machine and by the method described in [25, 26]. The specimens 2 mm in diameter and 15–20 mm in length were worn by friction against a silicon carbide grinding cloth, securely fixed to a revolving wheels. The specimens were pressed against the wheel under a load of 3 N, and while under load the specimens were moved over the revolving wheel from the periphery to the center. The tests were made under identical conditions for specimens of borided, carbided, and uncoated metals. The results of the tests were compared in the usual way by determination of the reduced linear wear for each specimen. Table 2 gives the mean values of the wear by weight and reduced linear wear of the tested specimens.

As can be seen from the table, the borided specimens are more wear resistant than the carbided specimens, both the borided and carbide diffusion coatings increasing by many times the resistance of refractory metals against abrasive wear. It should be noted that the weight wear and linear wear values of the borided and carbided specimens are rather high, since in some cases the thin diffusion layer was worn away before the test was terminated, and the further loss in weight occurred mainly as the result of wear of the metal itself.

Table 5. Results of Corrosion Tests of Carbided
and Borided Tungsten, Molybdenum, and Titanium
in the Cold for one Month (Weight before test, after
test, and weight lost, g)

Medium	Tungsten		Molybdenum		Titanium	
	carbided	borided	carbided	borided	carbided	borided
HNO$_3$ concentrated	2.5680 2.5676 0.0004	2.1641 2.1636 0.0005	4.1805 3.9230 0.2575	5.8832 5.8073 0.0759	0.5855 0.5754 0.0101	0.5302 0.5042 0.0260
HNO$_3$ (1 : 1)	2.7407 2.7407 0.0000	2.0743 2.0741 0.0002	—	—	—	—
H$_3$PO$_4$ concentrated	2.5389 2.5387 0.0002	2.0492 2.0487 0.0005	6.8523 6.8522 0.0001	6.1868 6.1868 0.0000	0.5442 0.5418 0.0024	0.5326 0.5120 0.0206
H$_3$PO$_4$ (1 : 1)	2.2630 2.2630 0.0000	2.1356 2.1342 0.0014	5.8568 5.8566 0.0002	6.1870 6.1857 0.0013	0.5228 0.5221 0.0007	0.5602 0.5518 0.0084
H$_2$SO$_4$ concentrated	2.9574 2.9573 0.0001	2.2500 2.2493 0.0007	5.6892 5.6892 0.0000	6.0540 6.0533 0.0007	0.5904 0.5382 0.0522	0.6024 0.5964 0.0060
H$_2$SO$_4$ (1 : 1)	2.5219 2.5218 0.0001	2.0709 2.0702 0.0007	6.6486 6.6483 0.0003	6.5912 6.5904 0.0008	0.6720 0.6429 0.0291	0.4322 0.4309 0.0013
KOH 30% 30%	2.0147 1.9971 0.0176	1.6171 1.5981 0.0190	5.7165 5.7164 0.0001	5.3338 5.3217 0.0121	0.5420 0.5402 0.0018	0.5328 0.5304 0.0024
KOH 10% 10%	1.9890 1.9412 0.0478	1.6026 1.5604 0.0422	6.2508 6.0094 0.414	6.0652 6.0150 0.0502	0.6245 0.6240 0.0005	0.5627 0.5624 0.0003

Note. Period of tests in KOH solution in the cold was 4 months.

Measurements of the microhardness of the diffusion coatings showed that it was close to the microhardness of the corresponding phases of stoichiometric composition. While having approximately the same hardness as the carbide layers, the boride layers protected the metal much more effectively from abrasive wear. Obviously, in assessing the ability of diffusion coatings to resist wear, it is necessary to take into account, in addition to their hardness, their brittleness, strength of bond with the metal base, and the structural characteristics, presence of pores, cracks, etc.

It was shown in [1, 10, 12] that the brittleness of borides is somewhat less than the brittleness of the carbides of the corresponding metals. In addition, bonding of the boride diffusion layers with the metal base is in all cases stronger than that of carbide coatings [13, 14, 21]. Evidently these factors also resulted in the better wear resistance of the borided specimens compared with the carbided specimens.

It is shown in [4, 6, 11, 25, 26] that there is a connection between the forces of the interatomic bond in the crystal lattice, which are characterized, for example, by the elastic modulus and the coefficient of lattice rigidity $m\Theta^2$ (m is the mass of a molecule of the compound, Θ is the characteristic temperature) and wear resistance. This connection was found in wear tests on the pure base metals and some alloys.

Currently, it is difficult to establish any correlation between wear resistance and bond strength in the crystal lattice in the case of refractory compounds, owing to the scantiness of e .perimental results of investigation of the wear resistance of these compounds, and the absence for many of them of the values of the elastic modulus, coefficient of rigidity of the lattice, and other values determining the force of the interatomic bond in the lattice. There is no doubt that further study of the wear resistance of refractory compounds (especially under conditions of abrasive wear) is of considerable practical and theoretical interest.

Testing for Chemical Resistance. The carbides and borides of refractory metals have high corrosion resistance in many inorganic acids and their mixtures [8, 13, 19]. Much work has been done by Russian and other scientists on the study of the resistance of these compounds in various chemically aggressive media; a complete bibliography on this question is to be found in the handbook [1]. The results of these investigations show that carbides and borides applied to metallic materials as coatings may considerably increase their corrosion resistance, and sometimes may also protect them completely from destruction. In the meantime, there is practically no information on the corrosion resistance of carbide and boride coatings and their protective properties.

We have carried out a series of experiments in which the corrosion resistance of borided and carbided specimens was tested in hydrochloric, sulfuric, nitric, and phosphoric acids (concentrated and diluted 1:1), and also in 10 and 30% solutions of caustic potash. The tests were carried out at room temperature for a lengthy period (20-30 days), and at boiling point for 1-3 hr. The corrosion resistance of the specimens was assessed according to their change in weight and external appearance after testing.

The results of the tests on the specimens are given in Tables 3, 4 and 5. Analysis of the data obtained has enabled several conclusions to be drawn. As can be seen from Table 5, carbided tungsten practically does not react with nitric, sulfuric, and phosphoric acids, and reacts only slightly with alkali solutions. Carbided molybdenum reacts slightly with phosphoric and sulfuric acids and alkali solutions, but dissolves rapidly in nitric acid. These results are in good agreement with the data of [4], in which an investigation was made into the corrosion resistance of powdered carbides of the transition metals.

Borided tungsten having on its surface a diffusion layer consisting of several phases reacted slightly with nitric, sulfuric, and phosphoric acids, and reacted fairly actively with alkali solutions. Borided molybdenum behaved similarly with the sole difference that it reacted much more actively with nitric acid. The results obtained are in agreement with the data of [9], in which the corrosion resistance of powdered borides of refractory transition metals was studied.

The analysis of the data on the corrosion resistance of carbided and borided specimens in boiling media (Tables 3 and 4) also confirms the results obtained in [4, 8]. Thus, carbided tungsten, molybdenum, titanium, and zirconium practically do not react with alkali solutions, but the reaction of the borided metals is quite considerable, especially in the case of tungsten and molybdenum. Similar results were obtained in [4] for the carbides W_2C and Mo_2C, and in [8] for the borided Mo_2B_5. Of the carbided specimens, tungsten with the phase W_2C and WC on its surface showed maximum chemical resistance, and of the borided specimens, niobium with the phase NbB_2 on its surface showed maximum resistance. According to [1], maximum corrosion resistance among the borides is shown by the diborides TaB_2 and NbB_2, and among the carbides, TaC, NbC, W_2C, and WC.

The results of the investigations carried out form the initial stage of the work which the authors intend to follow up in the future.

LITERATURE CITED

1. G. V. Samsonov ed., Analysis of Refractory Compounds, Metallurgizdat, Moscow (1962).
2. M. P. Asanova et al., Fiz. Metal. i. Metalloved. 9:689 (1960).
3. Yu. V. Grdina et al. (see present collection, p. 69).
4. I. V. Kashcheev and V. M. Glazkov, Izv. Vysshikh Uchebn. Zavedenii Fiz., No. 2 (1961).
5. V. P. Kopylova, Zh. Fiz. Khim. 34:1963 (1961).
6. V. D. Kuznetsov and V. I. Kashcheev Belorussk. SSR, Inz.-Fiz. Zh. Akad. Nauk 2:10 (1959).
7. G. A. Meerson et al., Sb. Nauchn. Tru. Min. Tsvet. Met. Zolot., No. 29 Metallurgizdat, Moscow (1958).
8. A. N. Minkevich, Metalloved. i Term. Obrabotka Metal., No. 8 (1961).
9. K. D. Modylevskaya and G. V. Samsonov, Ukr. Khim. Zh. 25:55 (1959).
10. V. S. Neshpor and G. V. Samsonov, Fiz. Metal. i Metalloved. 4:181 (1957).
11. I. V. Savitskii, Izv. Vysshikh Uchebn. Zavedenii Fiz. No. 2 (1958).
12. G. V. Samsonov and V. S. Neshpor, Collection, Question of Powder Metallurgy and the Strength of Materials, No. 5, Izd. Akad Nauk UkrSSR, Kiev (1958).
13. G. V. Samsonov and A. P. Épik, Coatings of High-Temperature Materials, Plenum Press, New York (1966).
14. G. V. Samsonov and A. P. Épik, Collection, Investigation of Steels and Alloys, Nauka, Moscow (1964).
15. G. V. Samsonov and Ya. S. Umanskii, Hard Compounds of Refractory Metals Metallurgizdat, Moscow (1964).
16. G. V. Samsonov et al., Boron Its Compounds and Alloys, Izd. Akad. Nauk UkrSSR, Kiev (1960).
17. G. V. Samsonov, Refractory Compounds, Metallurgizdat, Moscow (1963).
18. G. V. Samsonov and N. K. Golubeva, Zh. Fiz. Khim., 30:1258 (1956).
19. A. V. Smirnov and A. D. Nachinkov, Metalloved. i Term. Obrabotka Metal. No. 3 (1960).
20. Ya. S. Umanskii, Carbides of Hard Alloys, Metallurgizdat (1947).
21. A. P. Épik, Poroshkovaya Met., Akad. Nauk UkrSSR, No. 5 (1963).
22. M. M. Khrushchov and M. A. Babichev, Dokl. Akad. Nauk SSSR, 88:No. 3 (1953).
23. M. M. Khrushchov and M. A. Babichev, No. 9 in the collection, Friction and Wear in Machines, Izd. Akad. Nauk SSSR (1956).
24. M. M. Khrushchov and M. A. Babichev, No. 11 in the collection, Friction and Wear in Machines, Izd. Akad. Nauk SSSR (1956).
25. M. M. Khrushchov and M. A. Babichev, Dokl Akad. Nauk SSSR 131:6 (1960).
26. M. M. Khrushchov and M. A. Babichev, Vestn. Mashinostr., No. 3 (1964).
27. F. Gleser and W. Ivanik, J. Metals, 4:387 (1952).
28. A. Münster, Z. Elektrochem. 63:806 (1959).
29. E. Nikolaiski, Z. phys. Chem. 24:405 (1960).
30. H. Nowotny et al., Z. Metallkunde 50:417 (1959).
31. R. Steinitz et al., J. Metals 4:982 (1952).
32. W. Webb et al., J. Electrochem. Soc. 103:112 (1956).